Yaya Malo

Morphological and hydrographic mapping

Yaya Malo

Morphological and hydrographic mapping

sandstones of the Houet province (Burkina Faso)

ScienciaScripts

Imprint

Any brand names and product names mentioned in this book are subject to trademark, brand or patent protection and are trademarks or registered trademarks of their respective holders. The use of brand names, product names, common names, trade names, product descriptions etc. even without a particular marking in this work is in no way to be construed to mean that such names may be regarded as unrestricted in respect of trademark and brand protection legislation and could thus be used by anyone.

Cover image: www.ingimage.com

This book is a translation from the original published under ISBN 978-3-639-50772-0.

Publisher:
Sciencia Scripts
is a trademark of
Dodo Books Indian Ocean Ltd. and OmniScriptum S.R.L publishing group

120 High Road, East Finchley, London, N2 9ED, United Kingdom
Str. Armeneasca 28/1, office 1, Chisinau MD-2012, Republic of Moldova, Europe
Printed at: see last page
ISBN: 978-620-5-83698-9

Contents

ACKNOWLEDGEMENTS

This work bears our name, but altruism prevents us from taking credit for it alone. It is an inclusive work that has required the support of many people and institutions. It is therefore our duty to express our gratitude for their moral, financial and technical support, which enabled us to complete this study.

These are the same people who for years have been teaching us despite the always difficult working conditions. It is the teaching staff of the geography department of the University of Ouaga 1 Pr Joseph Ki ZERBO.

We make a special mention to our supervisor, **Pr Dapola Evariste Constant DA,** who has already supervised us in master's degree and who has accepted once again to accompany us in this new adventure. Thank you for your advice and guidance, so precious and indispensable.

We are grateful to Professor Madeleine KONKOBO/KABORE for her encouragement and support and to Desire POUSSOUGO.

We thank Mr Jean SIMPORE from the bottom of our hearts for his multiple supports.

We would also like to thank Desire KABORE and Remo KIENTEGA for their technical support and their contribution in satellite images respectively.

We are especially grateful to our field companions Anioue AKOBJORI and Bernadin SOMDA who agreed to climb the "cliffs" and hills with us while braving the harsh walking and access conditions.

We have a big thought for the students of the GIS and remote sensing laboratory (LT/SIG) of the geography department of the University Ouaga 1 Pr Joseph Ki ZERBO, as well as to Dramani Lamitou KOULANI.

Some institutional staff are not left out, such as Compaore Nestor from the Direction Regionale de l'Eau des Hauts-Bassins, Rene BAGORO from the BUMIGEB of Bobo-Dioulasso and KOROGO (IGB). Finally, we would like to thank all our guides without whom we would have had difficulty in surveying all these eminences.

SUMMARY

The sandy plateau of the Houet province (Hauts-Bassins region) constitutes the south-eastern edge of the Taoudëni sedimentary basin. This study on geomorphological mapping is carried out there. It is clear that cartography is nowadays a tool of choice in morphological studies in general and geomorphological studies in particular. The existing geomorphological maps of the geological plateau of the Houet province are small-scale and therefore insufficiently detailed. A large-scale map shows sufficient detail of the geomorphological units of the study area. The aim of this study was to produce a large-scale geomorphological map of the sandstone plateau of the Houet province.

For this purpose, it was necessary to interpret Landsat 8 satellite images of 30 m accuracy and radar (SRTM) images of 90 m accuracy. In addition, field verification of direct observations and the use of secondary data collected from a number of private and public institutions were indispensable. The collection tools consisted of a GPS, a data collection sheet and observation. Data processing required the use of GIS (ArcGIS 10.0) and remote sensing (ENVI4.5) software.

Thus, 14 geomorphological units could be defined. These are mainly plateaus, plains, lowlands and glacis. The most important geomorphological units are the sandy plateaus (35% of the area) and the least important are the middle glacis; valleys are also abundant in this province. The southern and south-western part is made up of plateaus and the north-eastern, extreme southern and south-eastern parts of the province are plains.

Key words: Sedimentary basin, geomorphology, mapping, remote sensing, valley, river.

GENERAL INTRODUCTION

The geological structure of Burkina Faso appears as a microscopic element attached to much larger structures, extending at least to West Africa from Algeria and Morocco to the Gulf of Benin (GEORGES H., 2002). These are birimian formations dating from the Paieoproterozoic era, i.e. 2.2-1.7 Ga (billions of years) of the West African craton, constituted by greenstone belts, delimited by granitoid batiolites of various natures (NABA S., 2007 ; ILBOUDO H., 2008). According to OUATTARA G., 1998, these formations are intersected by "*calc-alkaline and alkaline granite plutons and include the Man and Reguibat ridges and the Kenieba and Kayes "windows" (Map 1). The Birimian formations are bounded by mobile zones of Pan-African to Hercynian age, including the Dahomeyides and the Pharusian domain in the east, the Rockelides and Mauritanides in the west, and the Pan-African domain of the Moroccan Anti-Atlas in the north.*

However, a subsident marine basin individualises in the central part of this craton after it was strongly eroded and flattened during a calmer period: this is the Taoudeni Basin which extends over a large part of Mali, Mauritania, Guinea and a limited part of Burkina Faso (GEORGES H., 2002; YAMEOGO S., 2008), in which mainly sandy sediments were deposited between the Upper Precambrian (Precambrian A, from 1600 MA.) and the end of the Paleozoic (BESSOLES, 1977 in BOEGLIN J.-L., 1990). This basin separates the Man and Reguibat ridges (OUATTARA G., 1998; YAMEOGO S., 2008).

In Burkina Faso, sedimentary formations are poorly represented. According to NAKOLENDOUSSE S. (1991) and SAVADOGO A. N. (2012), 80% of the territory is occupied by crystalline or crystallopiyllian formations. These sedimentary formations, which represent 20% of all geological formations, are located in the western, south-western and northern borders of the country for the Taoudeni Basin and in the extreme east and south-east of the country for the Volta Basin according to GEORGES H., 2002; YAMEOGO S., 2008 and DA D. E. C., 2005. In other words, the sedimentary formations are less important. The Volta Basin covers Togo, Benin, Burkina Faso and Niger and the Gobnangou and Madjoari detritus formations are manifestations of these relics.

All these geological formations have evolved differently under the influence of several factors (physical and human) leaving varied and complex forms in the West African landscape. Therefore, GRIMAUD J.-L. (2014) finds that in non-orogenic areas, i.e. cratonic areas, climatic variations and epirogenic processes seem to control the topographic changes and the supply of sedimentary basins.

In Burkina Faso, these forms present themselves differently. The ëolian model in the north consists of dunes of decametrical heights often in east-west orientation, which covers, in places, granitic formations and often volcanic or sedimentary ones (DA D. E. C., 2005). In the centre and over most of the country, the differentially eroded crystalline basement granites provide a gently undulating peneplain landscape where the average elevation varies between 200-350 m. This landscape is also composed of domesticated areas, which are not very well adapted to the needs of the population. This landscape is also made up of granite domes and spines, voussoirs, tors and granite chaos. Then there are the volcanic formations, mostly birimian, consisting of chains, sometimes linear, sometimes curvilinear, such as the Kaya-Kongoussi arc, stretching over tens or even hundreds of kilometres (BOEGLIN J.-L., 1990; DA D. E. C., 2005). These alignments of convex hills sometimes take on the appearance of small "mountains". They are often covered with bauxitic or ferruginous armour, sometimes manganësiferes and surrounded by glacis layers, the first of which, armour, are often separated from their commanding reliefs by peripheral depressions. In the southern regions, the interfluves become shorter and their slopes steeper. The sedimentary formations in the west and north-west form a monotonous plateau with an altitude fluctuating between 450 and 600 m. The isolated eminences that emerge can exceed 700 m. Thus, the mountain of the classified forest of Beregadougou (717 m) and the highest point of the country, "mount" Tenan-Kourou (749 m), are in the horizontal system of the south-western gres of Burkina Faso (OUEDRAOGO C., 1983). They are in fact sandy flats according to SANOU D. C., 1984, on which differential erosion has carved an often ruiniform pattern of needles, domes, tables and anvils or mushroom shapes, which are clearly visible in the Sindou region with its peaks in the extreme south-west (DA D. E. C., 2005). The Banfora "cliff", representing an escarpment of more than 100 m, marks a south-east contact between the sandstone formations and the basement.

Towards the south-east of the country, other sedimentary formations belonging to a different sedimentary basin are marked by a generally clean rim dissected by the tributaries of the Pendjari, except in the area of the Gobnangou "cliff".

These sedimentary formations, although poorly represented in our country, are not sufficiently known and studied. With this in mind, we became interested in the geomorphology of the sedimentary formations, more precisely of the sandy plateau of the Houet province on the south-eastern edge of the Taoudeni Sedimentary Basin. The aim was to carry out a geomorphological mapping.

To achieve this, our study is divided into two main parts as follows:

-	a first part consisting of the theoretical framework and presentation of the study area and ;

-	a second part on the geomorphological mapping of the sandstone plateaus of the Houet province.

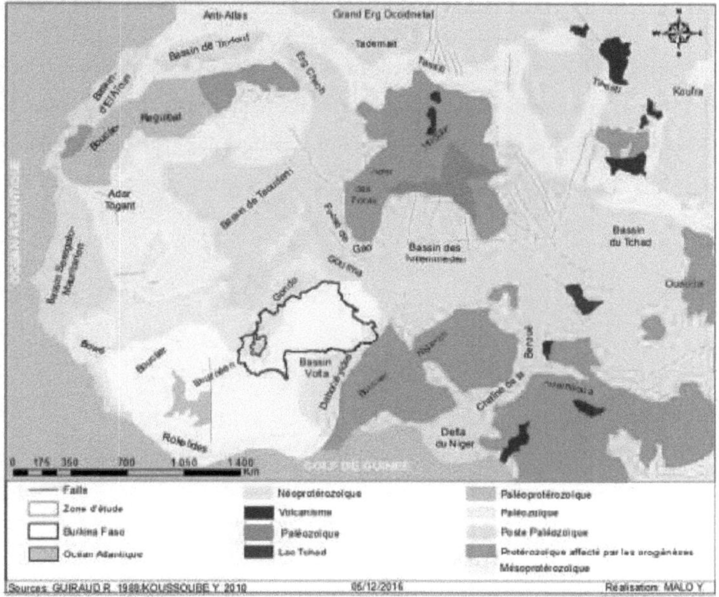

Map 1: The major stratigraphic subdivisions of West Africa

PART ONE

THE THEORETICAL FRAMEWORK AND PRESENTATION OF THE STUDY AREA

This part is based on the exploitation of secondary data, in particular documents of various kinds and sources, namely memories, articles and general works. It is divided into two chapters. The first chapter is essentially based on the methodology adopted and presents the study area, while the second describes the physical and human environment of the Houet province.

Chapter 1: Theoretical framework of the study
1.1. The problem

Today, the map appears to be an essential tool in many fields, including gëomorphology. According to TRICART J. (1965), maps and aërienne photographs constitute the basic documents in any morphological ëtude. These once prëpondërantes aërienne (PVA) shots made mapping almost inaccessible because of their rigidity. However, with the advent of satellites, the VAPs have gradually been relegated to the background. Indeed, located several hundred kilometres above the earth, these satellites provide sufficient distance to observe natural phenomena on the earth's surface (CHABREUIL M. et al., 1979). Coupled with the development of computer technology, these earth observation systems have made mapping more flexible and accessible.

Thus, many applications and uses are found for mapping; the fields are varied. These include weather forecasting, monitoring of agricultural resources, detection of mineral and oil resources, monitoring of sea and river pollution, management of water resources, monitoring of urbanisation and environmental changes, etc. (CHABREUIL M. et al., 1979; NARIMENE I., 2012).

As regards geology, hydrogeology and mining research, specialists in these fields such as: SAVADOGO A. N., 1984; ILBOUDO H., 2008; YAMEOGO S., 2008; CLENET H., 2009, have used remote sensing and mapping either to determine aquifers and their orientation, or to determine the deposits of certain minerals, the major geological units and their spatial distribution. It is very important to know that geomorphological phenomena are largely conditioned by the geological structures in place (BOEGLIN J.-L., 1990; GRIMAUD J.-L., 2014). For GEORGES H. (2002), it is not borrowing the geologist's own knowledge to question the nature of soil understanding, when the latter conditions the appearance, forms and evolution of the land. Clearly, it is difficult to study what is on the surface without knowing what is at depth.

In pedology, land use or vegetation, mapping plays a cardinal role in the distribution of the different soil types, their physico-chemical properties as well as their current dynamics which integrates the evolution of the soil cover (GUINKO S., 1984; FONTES J. and GUINKO S., 1995; YONKEU S., 2006; OUEDRAOGO I., 2010).

On the gëomorphological level, countries use cartography to study the current morphology of the relief or its dynamics. This is the case in the Aquitaine region of France where MALLET C. et al. (2007), have set up a decision support tool for coastal managers to prevent coastal erosion. The same is true for BILODEAU C. (2010), who used this tool to study the vëgëtation of the marshes in the Mont-Saint-Michel Bay. AVENARD J. M. (1977), also used cartography to relate the different elements of the physical environment to geomorphological units in the west of the Ivory Coast. In Burkina Faso, DA D. E. C., 1984 and SANOU D. C., 1984 used momentary mapping in their studies in the Gaoua and Kaya regions respectively, dealing with current dynamics. The observation is that geomorphological maps at the scale of a country are rarely made. The most available are those at the scale of a region, a coastline or a watershed.

Different users have different means and needs. Users use different types of satellite images and platforms. The processing tools, especially remote sensing and mapping software, are also diverse. In Burkina Faso, the Geographical Institute of Burkina (IGB) is in charge of producing administrative and topographic maps. Other specific services such as the Bureau des Mines et Geologie du Burkina (BUMIGEB) for geological and hydrogeological maps, the Bureau National des Sols (BUNASOLS) for the pedological map and sometimes the morpho-pedological map exist.

However, no structure is in charge of geomorphological mapping. Few geomorphological maps exist today in Burkina Faso and the first one was produced in 1989 by SANOU D. C. It was not until five years later, in 1994, that PETIT M., in collaboration with DA D. E. C. and GRANDIN G., refined this map to 1/1000 000 on the basis of the work of this author. This process was continued in 2005 by DA D. E. C. who increased the scale to 1/500 000.

While in many areas of Natural Resource Management (NRM), maps are constantly updated, for example in the field of the environment where the land use map has been updated every decade in Burkina Faso since 1992, the geomorphological map does not have such a fate and those available are small-scale and not sufficiently detailed. This is even more the case for the Houet province where there is a virtual absence of geomorphological maps. Those that do exist are sketchy with a high degree of generalisation. This study is initiated to make a large-scale geomorphological mapping (1/50 000) in the sandstone plateaus of the Houet province. In order to carry out this study, we asked ourselves a number of questions, the main one being: what is the contribution of 1:50,000 scale mapping in identifying geomorphological units in the sandstone plateaus of the Houet province? Two specific questions were also formulated:

- what is the impact of a large-scale map in identifying the different topographic irregularities of the sandstone plateaus of the Houet province?

- what types of geomorphological units might exist in the sandstone plateaus of the Houet province?

1.1.2. Research hypotheses

The general hypothesis

A 50,000 scale map[e] is an optimum means of identifying fairly well-detailed geomorphological units.

Secondary hypotheses

- a geomorphological map at 1:50,000 scale[e] of the sandstone plateaus of the Houet province provides details of the landforms;

- In addition to the plateaus and plains, there are other types of geomorphological units in the Houet province.

1.1.3. Research objectives

The main objective

The main objective is to produce a large-scale geomorphological map of the sandstone plateaus of the Houet province. This main objective is seconded by specific ones.

Secondary objectives

- To make a cartographic representation of the different gëoforms of the sandstone plateaus of the Houet province;

- identify the geomorphological units of the Houet province with their spatial distribution and importance.

1.2. The literature review

No study nowadays can claim to be new. In order to get acquainted with existing works related to our theme and our study area and to give a relevant orientation to our work, several documentation centres were visited in Ouagadougou and Bobo-Dioulasso. These included the libraries of the geography and history departments of the University of Ouaga I Pr Joseph KI ZERBO; the documentation centres of the BUMIGEB Bobo and the Direction Regionale de l'Eau des Hauts-Bassins; and the internet. Some private structures such as the Societe Africaine, d'Amenagement du Territoire (SAAT), a subsidiary of WATAM KAIZER and the Bureau d'Ingenieurs-Conseils (GERTEC) provided geotechnical data.

Thus, we were able to consult the work of researchers who have already used cartography and remote sensing in their research. It appears that in the South as well as in the North, cartography and remote sensing are used in the management and understanding of natural resources as well as in land use planning. The summaries of the work of some authors below confirm this.

DA D. E. C., YACOUBA H. and YONKEU S., 2008; KALOGA B., 1969; BETARD F. and BOURGEON G., 2009 have all used cartography as a scientific tool to improve geomorphological knowledge. In their respective works, they have linked pedology and geomorphology. The morpho-pedological maps produced represent units of the same name for which relief and soil are linked by strong interactions between geomorphological evolution and soil development. However, they have been less evocative of the structural aspect.

THOMAS Y. F., 1972: An inventory of forms and formations was made followed by a study of sedimentary budgets. He reviewed 34 systems of nomenclature of the granulometry of pebbles, sands and silts proposed since the end of the last century (1890) by researchers belonging to various disciplines such as geology, agronomy and public works. The different types of maps: frequency, synthetical, analytical and paramëtrical have been defined. As far as the legend elements are concerned, they are chosen according to the granulometry, coarse and medium sand, fine sand, silt, clay, gravel, sand. This study is concerned with the submerged environment and the shorelines. The forms have been described but there is no emphasis on geomorphological units. But the emphasis on sedimentology could be of great help in our study.

DOSSIN F. et al., 2009: their study is of great importance for our work because it effectively highlights the faults according to geomorphological units. The orientation as well as the folds of the faults are defined. These faults were of the underlain, undefined, thrust and hypothetical thrust types. It is a well-detailed map and the different components: hydrology, geology, especially aquifers, aquicludes, aquitards have been well represented and individualised. These are hydrogeological units. The karst system has also been mapped with a well-detailed hydrographic network where valleys, streams and valleys are represented. In addition to these maps, two thematic maps, a geological section and a hydrogeological section, and a lithostratigraphic table are presented.

GRIMAUD J.-L., 2014: studying the long-term dynamics of erosion in the cratonic context of West Africa since the Eocene, has not failed to make geomorphological, geological, topographical and simplified maps of the major tectonic and geographical units of Africa. In the north-west, the Adrar and Tagant plateaus in Mauritania are developed on sandy formations. In the south, these plateaus are separated from the Hodh depression by a major escarpment. This depression is covered by dunes and its rivers only join the Senegal drainage during very wet periods. To the south of this depression, the Senegal drains are perennial and cross the sandy plateaus of the Mandingo plateau and the Senegal-Mauritanian basin to the west. The Guinean ridge is composed of the Fouta Djalon in the northwest and the Simandou in the southeast. To the north-east of the Guinean ridge is the Haut-Niger area and the internal Niger delta. To the east, there are other sandy plateaus: the Banfora plateau and the Dogon plateau. The Gondo plain is located at the foot of the escarpment that delimits the latter to the east. Further east are located two low relief ridges which are outcrops of basement: those of Leo and Gourma. In the south, the Volta basin forms a depression surrounded by the sandstone plateaus to the west and the Dahomeyid reliefs to the east, both of which extend north-south. The Hoggar Massif in Algeria is prolonдё to the south by two apophyses which are the Air Massif and the l'Adrar des Ifoghas Massif. To the south of these massifs is the Bassin des luellemeden which is crossed by the lower course of the Niger. This study has enabled us to identify the great ensembles of Africa and to understand the geomorphological units of our country, namely their establishment and evolution.

LACAN C., 2010: he made a digital field model. Tertiary surfaces closer to the original Congo and Ogooue watersheds were reconstructed in order to determine the volumes of sëdiments deposited during this period. His aim was to find out what volume of formation was eroded during this period, and in particular the contribution of the Tertiary sands themselves. He then compared these volumes with those of today. This allowed him to perceive its evolution. It is clear that these forms have changed little since their emplacement. Most of the analyses were done with ArcGis software. A gëomorphological analysis and gëological sections were rëalisëes. His work could shed light on the methodology to be adopted for our study.

NARIMENE I., 2012: gëological mapping is of extreme importance for the gëologist, so it is essential to master all modern mëthodology and technique of making gëological maps. These include computer-assisted cartography, including tëlëdëtection and processing of satellite images. These geological mapping tools are the fastest, most accurate and most reliable available to geologists, especially when mapping in arid and desert areas where outcrops are often inaccessible, complicating conventional mapping. It was based on ENVI and GIS software and Landsat 7 ETM+ satellite images. Several combinations of bands were made and those optimal for identifying the different geological formations, lineaments, lithology were chosen. These combinations could be used to determine the geomorphological forms. The only difference is in the study areas: desert and Sudanian.

OKAINGNI J.-C., KOUAME K. F. and MARTIN A., 2010: working this time close to our study area, used remote sensing to map armour in the Ivory Coast. They proposed to apply the theory of belief functions (or Dempster-Shafer theory or evidence theory) to map breastplates. The presented approach consists in performing a classification using the theory of belief functions by merging information related to armour on images from neo channels obtained from the Landsat 7 satellite. Their contribution lies in the use of the k-nearest neighbour method and a fading coefficient to take into account, for each pixel to be classified, the uncertainties and imprecisions due to the presence of armour in the images. The method allowed the mapping of armour over the study area with a success rate of over 90%. The ENVI software was used and they took into account the vëgëtation index (NDVI) and the armour index (CI). This mëthod could be applied to our ëtude.

DAVEAU S., I960: it is the south-eastern part of the main mass of the gres that has been studied, in a region that stretches from Sikasso to Bobo-Dioulasso and Banfora. It attempted to establish the main features of the structure of the plateau and its border region and then to study their morphological aspects, controlled both by this structure and by the re-establishment of erosion surfaces, It was noted that two sectors of the plateau are still in the process of being eroded, and that the surface of the plateau is still being eroded by the erosion of the surface of the plateau. It has noticed that two sectors of different aspects oppose each other. From the Bobo-Dioulasso area to that of Bërëgadougou, the plateau ends along a remarkably rectilinear cliff, orientated from north-east to south-west, formed of sandstone dominating the bedrock by a continuous escarpment of 150 m in relative altitude. From Bërëgadougou onwards, the appearance of the cliff changes profoundly. The boundary between the sandstone and the bedrock turns to a general east-west and even south-east-north-west direction from the western Lëraba, and loses the remarkable straight line of the preceding sector. It has dëmontrëed that the sandstones are extremely varïëous. Her study mainly concerns the south of Bobo-Dioulasso to Sikasso in Mali. In this study we will consider the whole Bobo area.

DA D. E. C., 1989: He carried out a geomorphological mapping in the Centre-North region of Burkina Faso. It is a ëtude that was essentially based on mapping. Landsat TM 195-51 images of 18 January 1984 and 16 November 1986 were used. It is a 1:500,000 scale mapping[e] . The un^s identified could guide us in identifying the gëomorphological un^s of our study area.

1.3. The operating framework

It is represented in Table 1 below and shows how this methodology will be applied.

Table 1: The operating framework

Assumptions	Variables	Indicators	Materials
a geomorphological map at 1:50,000 scale• of the sandstone plateaus of the Houet province provides details of the landforms;	Unitesmorpho-structures/geoforms.	Eminence, depression, fault, lineaments, slopes	-GIS software (ArcGis 10.0) ; -Remote sensing software (ENVI, PCI Geomatica V9.1, RockWork 15); -satellite image ; -radar image
In addition to the plateaus and plains, other types of geomorphological units exist in the Houet province.	Geomorphological units	Plateau; glacis; plain; cliffs; lowland	-GIS software (ArcGis 10.0) ; -Remote sensing software (ENVI, PCI Geomatica V9.1, RockWork 15); -satellite image ; -radar image.

1.4. Clarification of concepts

CARTOGRAPHY

Cartography is defined by GEORGE P. and VERGER F. (2009), as a set of techniques leading to the production of maps. The use of aerial photography associated with photogrammetry, satellite tëlëdëtection and computer technology has led to a complete renewal of the modes of production and updating of most types of maps. It is defined by the Societe Frangaise de Photographie et de Teledetection (1989), as a "computer system allowing, from various sources, to gather and organise, manage, analyse and combine, elaborate and present geographically localised information, contributing in particular to spatial management". In other words, GIS enables the processing, management, analysis, integration and modelling of geographical data, as well as the processes that transform the territory. In this respect, it is necessary to manage integrated, multidimensional information representing complex environments, in addition to being able to model development scenarios.

GEOMORPHOLOGY

Geomorphology is defined by **GAUDIN S. (1996)** as the study of relief, whether at the descriptive or interpretation level. This discipline involves notions of stratigraphy (number of layers, thickness, angular relationship...), lithology (surface states), tectonics (study of folds, faults...). It is also based on the knowledge of erosion phenomena which depend not only on the rocks in place and their alteration, but also on climatic conditions.

BADOUX H. (1989), defines gëomorphology as the study of the model, to which various agents can contribute: wind, snow, glaciers, dissolution, runoff and rivers.

SANOU D. C. (2008), defines gëomorphology as a science of earth forms. Its purpose is to observe, describe and explain the forms of the earth's relief. It is subdivided into three branches according to SANOU D. C., which are :

- **structural geomorphology**: which is the study of the structure, that which does not move, that is to say, the nature of the various rocks, their dispositions and the general forms which are characteristic of them;

- **climatic geomorphology**: **the** aim of which is to distinguish different areas in each of which the modalities of erosion are different and in each of which the shape of the worn rocks will present a common air for a common agent. In other words, it is the science which studies the relief in its relationship with the bioclimatic environment;

- **dynamic geomorphology**: which is a science that aims to shed light on the modalites, the dynamism of the earth's landforms.

For **MASSON M. (1972)**, geomorphology is a science which is interested in the forms of the current terrestrial relief and their evolution, which depends on :

- physical and chemical characteristics of rocks ;

- climate, vegetation, altitude (external geodynamics) ;

- the action of internal geodynamic forces (seismicity-volcanism-orogeny) ;
- of human action (anthropogenic phenomena) ;
- the interaction of different factors with each other.

For our part, it is a science that studies the different forms on the surface of the earth, taking into account the substratum in place and the exogenous and endogenous factors.

GRESEUX PLATEAU

The plateau is a geographical area of varying elevation, where the watercourses are deep, in contrast to the plains. The limits of the plateau are areas of change in relief or altitude, they can be marked by steep escarpments or slopes, these spaces are called slopes in topography. http://littre.reverso.net/dictionnaire- francai s/definition/gres/3 5776 -10/09/2016.

The term "sandstone plateau" is used when the plateaus are formed on sedimentary formations composed of sandstone. The term *"sandstone"* refers to *a cemented assembly of grains of relatively homogeneous grain size composition. The size of the grains varies between 2 cm and 50 u (0.05 mm) in diameter. According to the nature of the cement, we distinguish essentially: siliceous, calcareous, ferruginous gres"* (SANOU D. C., 2008). For our part, a sandstone plateau is an eminence or a group of eminences of medium height formed on ancient sedimentary basins. These eminences have been modelled by erosion, especially differential erosion.

SEDIMENT BASIN

There are several definitions of sedimentary basins:

According to https://fr.vikidia.org/wiki/Bassin sedimentary (10/09/2016), a sedimentary basin is: *"a large natural region whose basement structure is composed of superimposed layers of rocks. They were formed in basins of the earth's basement generally covered by water. The rivers coming from the continents pour debris that erosion has torn away from the emerged reliefs. This debris of various origins (mineral, animal, vegetable) which has been deposited in seas, lakes or lagoons, or in any depression, becomes sediment.*

The extent of the layers is related to the extent of the troughs which have experienced, in the course of geological time, marine invasions (transgressions) but also marine reflux (regressions). Two successive layers therefore do not have the same size or the same limits. The thickness of the sedimentary layers is variable. It is related to the amount of debris input, but also to the time taken for the debris to accumulate. The weight of the debris slowly sinks the bedrock, which may allow considerable accumulation as is the case in the southern Mississippi plains, where the bedrock is more than 8,000 metres below the sedimentary soil surface.

GEORGE P. and VERGER F., 2009 define it as "an area of subsident marine sedimentation, generally shallow, located on the edge or within a craton. The sediments pile up in shallow layers that may form aureoles from the përiphërie towards the centre (regressive sedimentation). The dëp6ts are very varied and often correspond to a sedimentary cycle."

COQUE R. *in* Encyclopedia Universalis (online), consulted on 10/09/2016 has defined the sedimentary basin as *a more or less regular bowl-shaped geomorphological unit, characterised by a combination of specific structural forms, flattening surfaces and accumulation forms. Sometimes, the basin arrangement is expressed by a certain convergence of the hydrographic traces. In these cases, the sedimentary basins are named after the great rivers that benefit from them (Congo, Zambezi, Amazon, Mississippi). We choose this last definition which we consider more geomorphological in the framework of this study.*

TELEDETECTION

Tele means "at a distance" and detection means "to discover" or "to detect". The term "remote sensing" first appeared in the United States in the 1960s, when new sensors were added to traditional aerial photography. The term teledetection was officially introduced into the French language in 1973 and its official definition is as follows: "All the knowledge and techniques used to determine the physical and biological characteristics of objects by means of measurements carried out at a distance, without physical contact with them" (Commission interministerielle de terminologie de la teledetection aerospatiale, 1988).

According to CHABREUIL M. et A. (1979), the term teledetection refers to a set of techniques whose purpose is to study either the surface or the atmosphere at an altitude of between a few metres and several thousand kilometres. It requires sensors to collect the information and vehicles in which these sensors (passive or active) are carried. A receiving station is also essential. Finally, various processing systems are needed to make the data collected usable by earth scientists.

VALLEY

According to the dictionary Le Petit Larousse illustre, 1992, a valley is an elongated depression more or less evasive, shaped by a river or a glacier. For the geomorphologist, the bottom of a valley constitutes a talweg. The highest parts of the slopes (ridge lines) constitute water demarcation lines and limit the watershed. The river usually has a minor and a major bed which are often flood areas. Valleys usually have a permanent or temporary watercourse, otherwise they are called dry valleys (https://fr.wikipedia.org/wiki/Vall 11/07/2017).

WATERCOURSES

A watercourse is the general name for flowing waters that circulate through a fixed channel. They are named according to characteristics of flow, size..., such as rivers, streams, gullies, brooks, torrents... While the majority of rivers are permanently visible at the surface, some are underground and others are temporary https://www.aquaportail.com/definition-3804-cours-d-eau.html 11/07/2017. **In hydrology**, a river is a collection of water fed by springs, groundwater and runoff from precipitation. From the source to the mouth, the living conditions in rivers are constantly changing and, along with these changes in the environment, a variety of flora and fauna succeed one another. Rivers transport continental waters over their beds and between their banks.

Geographers, on the other hand, are accustomed to distinguish between **rivers**, which flow into the sea through an estuary or delta mouth, and streams, which flow at a point called a confluence into a river or into another river, or into a lake. In addition, when the watercourse has an impetuous flow, and its slope exceeds, on average, 0.05 m to 0.06 m per metre, it is more specifically called a **torrent**. If its bed is not very wide and its flow minimal, it is a **stream** or a **ru**.

1.5. Methodology and presentation of the study area

1.5.1. Preliminary work

The first step consisted of the digitisation of the hydrogeological map at 1:500,000^e of the BUMIGEB and the Geological map of Burkina at 1:1,000,000^e according to HOTTIN G and OUEDRAOGO O. F. (1975) and the map of the major geological units of West Africa. We were thus able to delimit the study area and situate it in relation to West Africa and then in relation to Burkina Faso. The Bobo-Dioulasso sheet of the 1:200 000 geological map^e was used. A memorandum of understanding between the remote sensing and GIS laboratory (LT/GIS) and the Geographic Institute of Burkina Faso (IGB) provided data in shape files at 1:200,000^e on topography, geology, soils, hydrographic network, habitat and vegetation.

1.5.2. The material used

The equipment consists of a GPS, a camera, a GPS data collection sheet. In addition, the computer equipment consists of GIS software (ArcGIS 10.0) and remote sensing software (PCI Geomatica V9.1, ENVI4.5, RockWork15).

1.5.3. Image and map data acquisition

Two types of images were required for this study. These are mainly Landsat 8 satellite images and the Digital Terrain Model (DTM). For the latter, the 90m resolution Shuttle Radar Topography Mission (SRTM) image downloaded from the West Africa Science Service Center on Climate Change and Adapted Land Use (WASCAL) geoportal was used.

1.5.4. Data collection

1.5.4.1. Geomorphological data

One of the preferred techniques in geomorphology is the description of the different patterns that can be observed in the landscape. It is a complete inventory of the observational data concerning the relief and those necessary for its explanation. The inventory takes into account topography and structure with an explanatory aim: to understand the formation of relief. Thus the following data have been collected:

- **morphographic (purely descriptive)**: correctly identify landforms and forms of dissection, ablation and accumulation;

- **morphometrics**: assessment of the extent of gradients and the value of slopes. These data are incorporated into the symbols or expressed as significant contours;

- **structural (explanatory)**: relationship between the relief in place and the nature of the rocks. It is a selection of important data from the geological map;

- **morphogenics:** account for the conditions under which landforms are formed, taking into account erosion factors;

- **chronological:** succession in time of the different generations of forms. It is a question of their arrangement or nesting while specifying their age.

1.5.4.2. The realisation of transects

Three axes of 30 to 95 km in length have been realised:

- axis 1: Bobo-Dioulasso-KotUdougou passing Yëguëresso towards the east on the national road n°1, 30 km long;

- axis 2: Bobo-Dioulasso-Toussiana via Darsalami, Pëɯ over a distance of 55 km;

- axis 3: Bobo-Dioulasso-Fo via Bama, Samandeni, Dandë, Koundougou, 95 km long.

The objective of these transects is to cover all 9 sandstone formations in the province. The sites are reached by motorbike and the rest by foot with the help of guides.

1.5.4.3. Data processing

Secondary and primary data processing was done with mapping and tëdëtection software. ENVI4.5, ArcGis 10.0, Geomatica V9.1, RockWorks15 were used for Landsat 8 and SRTM image processing. ArcGis 10 for mapping. The transfer of data from the GPS to the computer was done with software such as MapSource and DNR GPS. Excel and Word 2010 for tables, graphs and text entry.

1.5.5. Presentation of the study area

This study takes place in the province of Houet, one of the three provinces of the Rëgion des Hauts-Bassins. It is home to the regional capital, Bobo-Dioulasso, located 365 km from the capital, Ouagadougou. Created on 15 September 1983, the province covers an area of 11,540 km^2 and represents 4.21% of the national territory. Administratively, it has 205 villages and 13 rural communes/departments. It is composed of the communes of Bama, Dandë, Faramana, Fo, Karankasso-ViguU, Karankasso-Sambla, Koundougou, Lena, Padema, Peni, Satiri, Toussiana and Bobo-Dioulasso. Five main axes cross the province: RN 1 and 11 to the east, which lead respectively to Ouagadougou and Diëbougou-Frontiëre from the Republic of Ghana; RN 10 to the northeast, which connects Bobo to Dëdougou; RN 7 to the southwest, which connects Bobo-Dioulasso to the Republic of Côte d'Ivoire via Banfora; the RN 8 to the west which carries traffic between Bobo-Orodara-Frontiëre with the Rëpublique du Mali and the RN 9 which links Bobo to Faramana-Frontiëre with the Rëpublique du Mali. The province is also crossed by an Abidjan-Ouagadougou railway in the southwest/northeast direction.

The province of Houet is bordered to the north by the République du Mali and the province of Banwa, to the south by the province of Comoe, to the east and north-east by the provinces of Tuy and Mouhoun, to the south-east by the province of Bougouriba, and to the west by that of Kenedougou.

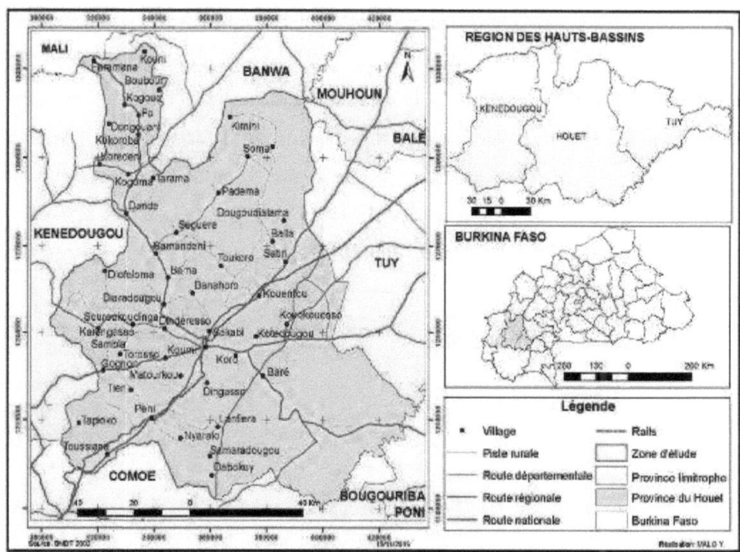

Map 2: Location of the study area

Chapter 2: The physical and human framework of the Houet province

The province of Houet remains one of the most impressive in Burkina, both for its physical features and its human composition. It also offers several assets for the development of socio-economic activities.

2.1. The physical setting

The physical environment is defined as the climate, the water system, the soil, the vegetation, the relief, in short everything that is natural without any form of human intervention.

2.1.1. The climate

The climate is Sudanian and influenced by humid air masses from the coasts and dry, warm air masses from the Sahara. When the inter-tropical front (ITF) moves southwards, the dry season lasts five to six months (November/December to April) and when it moves northwards, the wet season sets in and lasts six to seven months (May to October/November). Climate remains one of the major factors influencing the geomorphology of a region apart from telluric activities (tectonic movements and volcanism). It acts through factors such as temperature, precipitation, winds, insolation, air humidity, evapotranspiration...

> **Rainfall in the Houet province**

A series of rainfall data from the Bobo meteorological station from 1950 to 2013 shows that annual rainfall varies between 775 mm in 1977 and 2011 making them the driest years to 1552 mm in 1952 which is the wettest year. Thus, there is a significant inter-annual irregularity in rainfall. The interval of variation calculated according to the ratio between the rainfall of the wettest year and that of the year with the lowest rainfall (1552 mm / 775 mm) of AVENARD J. M. (1977) is equal to 2. This ratio therefore shows that there are disparities which are often hidden by the average. The monthly averages for this series are shown in Graph 1 below. July, August and September are the wettest months with about 200-300 mm of rain.

Figure 1: Average monthly rainfall from 1950-2013

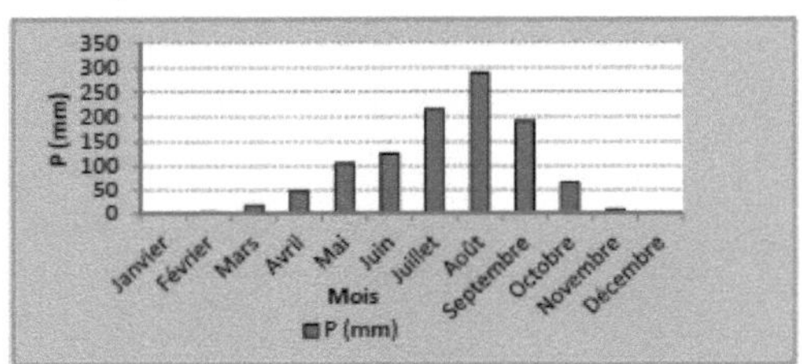

Sources: Bobo-Dioulasso meteorological station (1950-2013)

> **Temperatures**

The sërature data from the Bobo-Dioulasso weather station from 1951 to 2013 has shown that March and April are the warmest months with maxima around 35, 36 and 37 °C and minima around 19-24 °C. During July, August and September, the maximums fall to 30 °C and the minimums to 18-19 °C, on average, in December and January. The average annual temperature range, which is the algebraic difference between the average temperature of the warmest and coldest month, is 18.45°C. Temperatures can weaken rocks, exposing them to erosion.

Graph 2: Average monthly temperature variations from 1951-2013

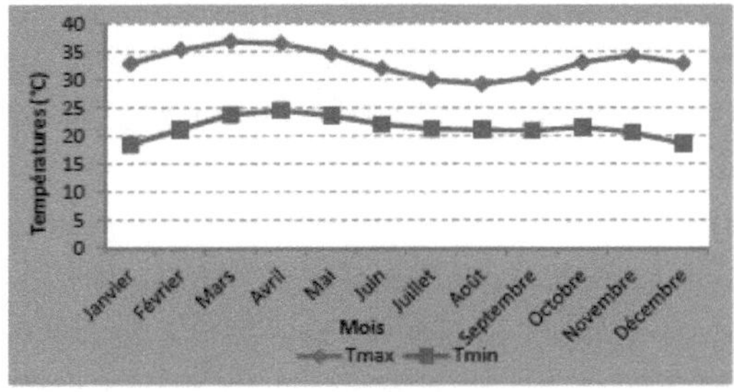

Sources: Bobo-Dioulasso weather station

__Tmax__: maximum temperature and __Tmin__ (minimum temperature)

Sunstroke

The average monthly duration of insolation from 1950 to 2013 shows that July, August and September are the least sunny months with an average sunshine of less than 200 hours. November, December and January are the sunniest months with an average sunshine duration above or equal to 250 hours. This sunshine duration as well as the temperature affects the geological substratum in place.

Figure 3: Average monthly insolation from 1950-2013

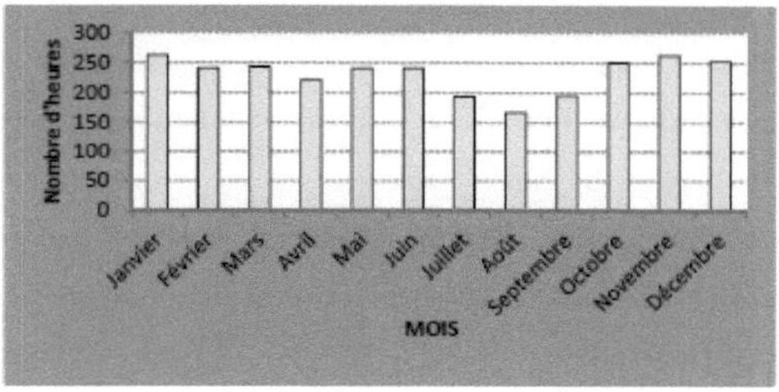

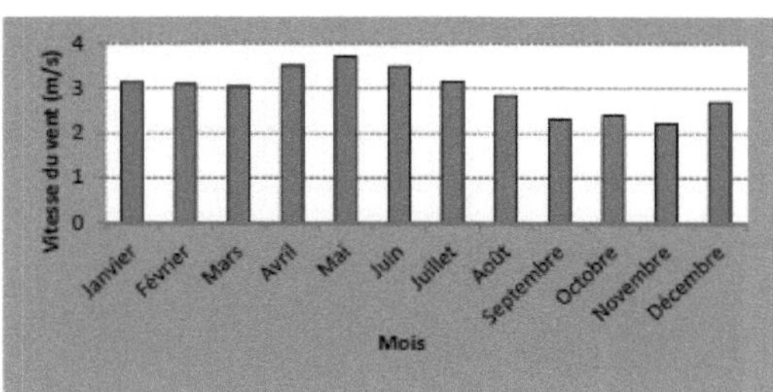

Source: Bobo-Dioulasso weather station

The winds

The monthly averages of wind speed for the 1961-2013 period show that for some months they blow at more than 3 m/s on average and for other months at a little less. This is the case for the months of August, September, October, November and December where this speed is low. During the months of May, April and June, the winds blow quite strongly, thereby extending the rainy season. The wind is an agent that acts on morphology in the sense that it can destroy or build a geomorphological unit.

Figure 4: Wind speed from 1950-2013

Source: Bobo-Dioulasso weather station

2.1.2. Flora and fauna

The province of Houet belongs to the South Sudanese phytogëographic domain according to the dëcoupage of FONTES J. and GUINKO S. (1995). It is a vëgëtation of savannahs in all its subtypes, i.e. from the wooded savannah in the south, to the shrubby savannah in the north,

23

passing through the shrubby and man-made savannahs or pare. The open forest is observed in the south of the province with a stratum that lies between 15-20 m and contains species such as *Afzelia africain, Daniela oliveri, Kaya senegalensis, Acacia seyal.* In addition to these savannah formations, there is a lush vegetation formation along the watercourses when untouched by man. These are gallery forests with a stratum between 15-25 m. The plant species in this forest are: *Antiaris africain, Berlinia grandifolia, Mitragyna inermis, Anogeusus leiocarpus.*

All these formations provide a habitat for wildlife which is important in the province. This fauna consists of big game such as elephants, buffaloes, hippotragues (coba), waterburcks, hippos, hartebeests, buffon cobs, redunca, harnessed guibs and small game consisting of phacocheres, porcupines, ewes, monkeys, hares, monitors, rats, snakes, herps, aulacodes.

2.1.3. The geomorphology of the Houet province

The Houet province has an uneven relief in its central, southern and northern parts. The peneplain on the bedrock and the sandy plateaus form the two morpho-geological complexes of the province (map 3). These are :

- of the Precambrian peneplane basement which occupies the eastern part and extends from Koro to Hounde for 80 km. It is marked by lateritic glacis or cuirasses, softly undulating and low-lying with a barely marked hydrographic network. The average altitude varies between 250 and 300 m;

- This is part of the Precambrian A sandstone sediments which form a monotonous relief marked by the Banfora "cliff" which dominates the crystalline peneplain by 100 to 300 m (IWACO, 1993).

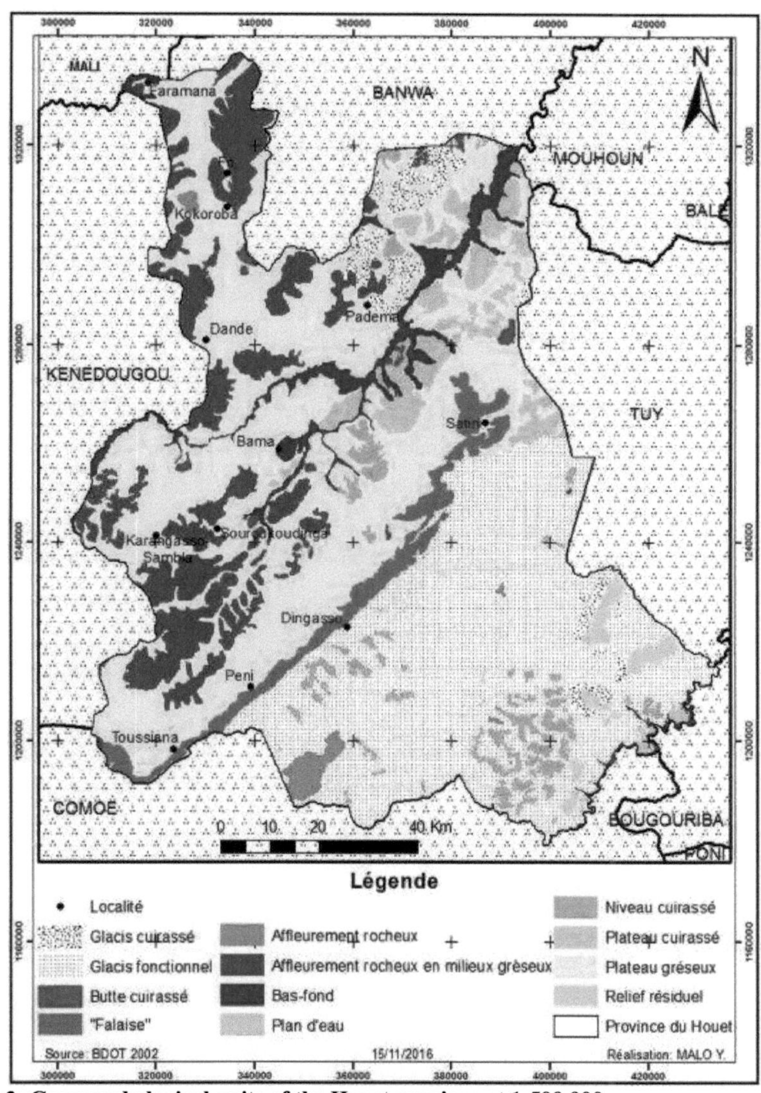

Map 3: Geomorphological units of the Houet province at 1:500 000

2.1.4. The river system

The Houet province has a fairly dense hydrographic network with important rivers. The most important river of Burkina Faso has its source in the sandy plateau in the south-west, 70 km from Bobo-Dioulasso (BICABA K. et al., 1998). It is the Mouhoun (ex Volta Noire). The rivers that drain the waters of the Houet province are divided into three major international river basins, namely

- The Niger River is fed mainly by the Banifin and its tributary Tessë composed of Seledogo and Sesse which flows northwest to join the Niger River;
- The Volta basin, more precisely the Black Volta, which occupies almost the entire province and is fed by Leysso, Bougouriba, Grand Bale, Ouolo, Ouere, Tionou, Mou, Mouhoun, Koba, Siou, Kou, Dienkoa, Kiene. The upper Mouhoun flows northeast, then branches off into the Sourou and heads south where it becomes the lower Mouhoun before flowing into the Atlantic Ocean in Ghana after a confluence with the Nazinon and Nakambe at Lake Volta and ;
- The Comoe-Leraba basin is fed by Sinlo in the extreme southwest. The Comoe River rises in the sandy plateau near Peni and is joined by the Leraba, one of its main tributaries. These rivers flow southwards into the Atlantic Ocean via the Ivory Coast.

There are significant flood zones in the province, mainly in the Mouhoun basin (Map 4).

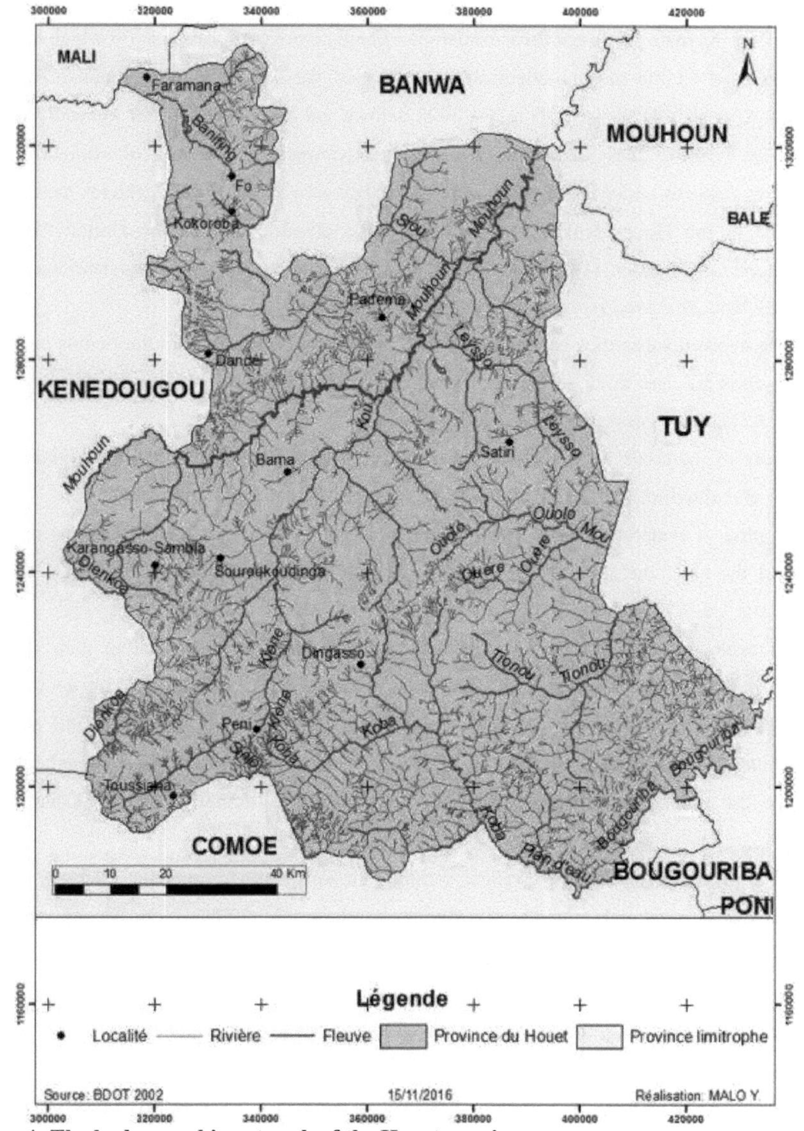

Map 4: The hydrographic network of the Houet province

27

2.1.5. Soils

Soils reflect the main factors of their formation, climate, topography, vëgëtation which in turn is conditioned by the physico-chemical properties of the soil, time and human influence (JONES A. et al., 2015). SANOU D. C., 2008 defines soil as *"a more or less soft and friable superficial layer of firm earth resulting from the desaggregation and alteration of the underlying bedrock under the effect of physico-biological agents"*. Six soil classes have been identified in the Houet province, according to the classification of the United Nations Agriculture Organisation (FAO) and adopted by the national soil office of Burkina Faso (BUNASOLS, 2003) and represented on Map 5. These are

- **little evolved soils:** they have a profile that is not very different from the humus horizon (A), which is not very thick, to the original material, which is more or less altered (C) or not altered (R). Two processes define this group of soils:

the supply of materials by runoff, either laterally from high to low levels (colluvium), or along rivers (alluvium). These have undergone little pedogenetic evolution;

the stripping of materials in situ by water erosion, the intensity of which depends on the nature of the rock, the slope of the environment, the surface condition of the soil, human factors, etc.

Poorly developed soils have a low chemical water holding capacity due to their coarse texture, limited thickness and runoff losses. Their chemical fertility depends on the geological nature of the bedrock, but is generally low. These soils are used for millet and peanut cultivation and can also be used as a grazing area for livestock. They are found throughout the province but remain important in the north and centre, particularly towards Kouni and Lanfiera;

- **Fersiallitic soils:** these are tropical ferruginous soils with little or no leaching. They occupy a band that extends from the centre to the south-west of the province. Because of their low chemical content, their exploitation must be accompanied by the addition of organic manure. They are used to grow cereals (millet, maize, sorghum), groundnuts and cotton. They are on the glacis;

- **hydromorphic soils:** these are soils with little moisture and a pseudogley surface. These soils occupy the drainage network and evolve under the influence of excess water. They are deep (above 100 cm) with poor drainage. This class of soils is suitable for rainfed and rainfed rice crops and market gardening. They are mainly located along the Mouhoun River and the Bougouriba River, as well as in secondary streams and lowlands;

- **Crude mineral soils:** these are skeletal soils with low organic matter and are characterised by a lack of pëdological evolution. The class of raw mineral soils is subdivided into two subclasses such as Lithosols on armour (L/c) and Lithosols on rock (L/r). Crude mineral soils lack a sufficient base for root development. Their agronomic suitability is therefore low to nil. They are less abundant in the province and are located in the extreme northwest near the Banwa province. They also occupy the glacis;

- **mull soils**: belong to the group of eutrophic brown soils. They are associated with volcano-sedimentary formations. Their composition is sandy-clay on the surface and clayey at depth. They are mainly found in the south-east of the province;

- **soils with sesquioxide and rapidly mineralised organic matter**: they belong to the group of leached tropical soils. These are soils with very extensive primary mineral alteration, which generally takes place at depth, while the surface organic matter undergoes a very rapid evolution. They are the most abundant in the province.

These different types of soils are rapidly depleted and require chemical fertilisers for their agricultural use. They also show a high sensitivity to erosion. They evolve according to geomorphological units.

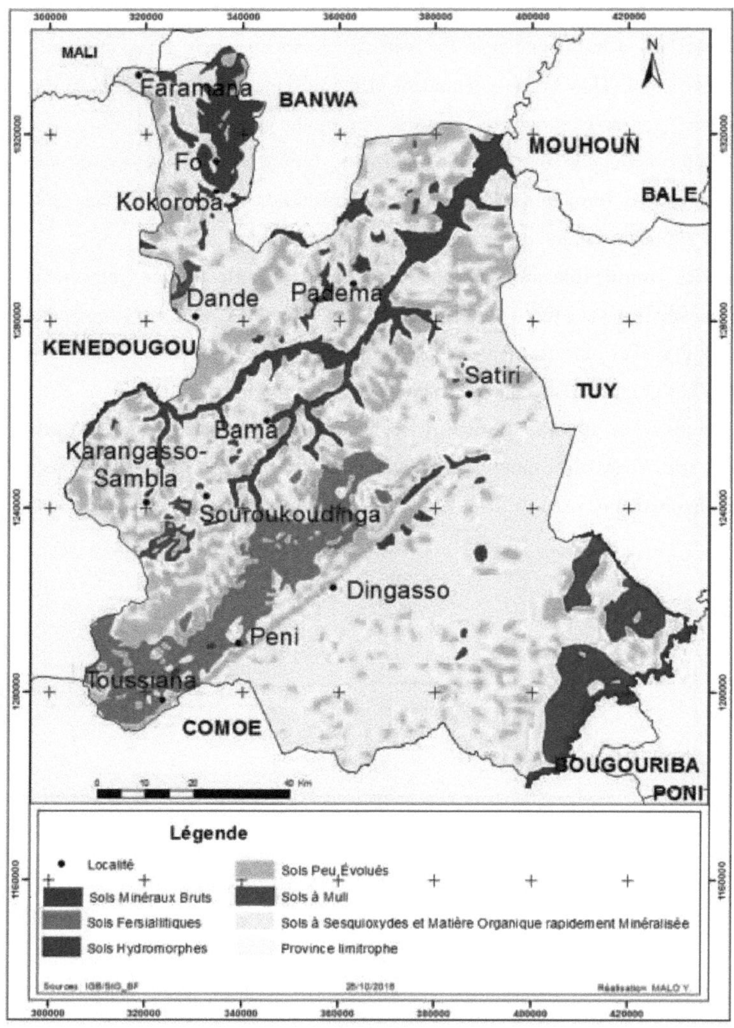

Map 5: Soil classes of the Houet province

2.1.6. The geology of the Houet province

The Houet province is on the contact zone of the two major geological domains of West Africa, namely the Precambrian basement on the one hand and the primaries or paleozoic covers on the other (DAVEAU S., 1960). The works of SANOU D. C., 1984 and those of IWACO (1993) give sufficient details of the geological formations of the province. OUEDRAOGO C. (1983) and DAKOURE D. (1999) counted nine types of sedimentary formations in the Houet province, all of them Upper Precambrian. NABA S. (2007) defines well the different gëological periods during which these formations were established.

2.1.6.1.　　Base courses

The Antebirimian or Precambrian D or Lower (2.4-2.15 Ga): these formations are essentially granitoids, namely granites and migmatites, gneisses. These gneisses appeared following folding and metamorphism of the granitoids (SANOU D. C., 1984) during the Liberian orogeny (2.8-2.7 Ga). They occupy the eastern part of the province.

The Birimian or Precambrian C or Middle (1.3 Ga): the Birimian series consists of a variety of rocks of volcanic, volcanic-sedimentary, clayey-sedimentary and pyroclastic origin. They were emplaced in elongated trenches or basins (20 to 50 km wide and 100 to 500 km in extent).

2.1.6.2.　　Chronostratigraphy of the sedimentary cover (Precambrian A)

The Precambrian A deposits are mainly sandy interbedded with shales or, more rarely, calcarodolomitic. The maximum height exceeds 1500 m. From bottom to top, nine sedimentary formations can be distinguished:

- **the lower layers**

These gres are very weakly represented in the southern part of the province. From the base to the top, fine conglomerated arkosic sandstones and fine to medium, white or greenish quartzites are observed, with a pass of coarse conglomerated feldspathic sandstones, pink sandstones with schistose debits and finally, a schisto-gresous level which is often very altered. Its thickness is estimated at 300 m.

- **basic gres (Kawara-Sindou)**

These rocks lie unconformably on the crystalline or crystallophyllous basement. Their thickness varies between 60 and 350 m. They extend from the south-west to the north-east where they are important and reach their maximum power and extension at outcrop. The formation is subdivided into two facies:

- the lower part has a variable granulometry. In the west, coarse conglomerates become finer towards the top; towards the east, fine conglomerates become increasingly coarse at the top with lenticular conglomerates with large decimetric pebbles.

31

- the upper part is made up of fine grey-white, well-graded sandstone in thin centimetric banks.

- **fine glauconitic gres or Sotuba gres**

They lie on top of the basic sandstone and extend from south-west to north-east with a thickness of 300-500m. This formation shows an alternation of coarse conglomerated glauconitic sandstone and thin levels of very fine, silty, reddish sandstone with a schistose finish, characterising its base. This coarse base is about twenty metres thick and disappears in places. Above this, there is a level of greenish clay with a schistose flow that can be up to 80 m thick. The colour is related to the evolution of the alteration.

- **quartz pebble stones**

They lie on top of the Sotuba rocks and are larger than these. Their thickness varies between 0 and 500 m. They are heterogeneous and comprise fine to coarse quartzite beds with criss-crossing stratification. Schistose and quartzite intercalations have been observed towards the top of these beds, which extend from the Bobo area in the form of armoured plateaus to Orodara. They have a south-west to north-east direction.

- **the greso-schisto-dolomitic or Guena-Souroukoudinga stages**

They are underlain by quartz pebbles and begin with coarse reddish-brown, glauconitic gres, containing thin intercalations of very fine, pink, well bedded, micaceous gres. On the upper part of the facies there are intercalations of mauve dolomite and greenish argillite. This facies is surmounted by large decimetric to metric banks of very fine, arkosic, ash-grey to pink, finely Шё and ткасё in the beds and sometimes glauconitic gres-quartzite. Next comes an alternation of greenish, micacёe, gresous argillite with schistose dёbit and reddish siltstone in platelets. This alternation of argillite with siltstones seems to constitute the rest of the formation, with in places, large banks of very fine sandstone or more or less thick levels of dolomite and dolomitic limestone with stromatolites. The total thickness of the formation could exceed 300 m.

- **the pink gres**

They are underlain by greso-schisto-dolomitic strata and occupy the whole western part of the province, heading northwards. These essentially quartzic deposits are characterised by homogenous fine and compact levels. Their colour is variable, generally pink and sometimes greyish. Rare and small intercalations of dolomitic breccias are observed at the bottom of the formation.

- **The Samandeni-Kiebani siltstones, argillites and carbonates or Toun shales**

They are located in the northwest of the province and slightly to the northeast. This formation includes shaly, ferruginous and shaly gres as well as dolomitic passёes. It is very arable, with few outcrops, and is found in the northwest and west of the province.

- **the gres of Koutiala**

The lower part consists of alternating silty claystone and glauconitic siltstone containing thin banks of dolomitic limestone with stromatolites. The remainder of the formation appears to correspond to well-bedded, micaceous, shaly-bedded green claystone containing levels of dolomitic limestone, granular limestone often completely silicified, flint and rare banks of very fine, silty sandstone. The thickness of this formation is estimated at about 450 m. They are sparsely represented in the northwest of the province.

- **the Bandiagara gres**

They constitute the terminal stage of the Precambrian A formations and are found in the Fo region where they overlie the Koutiala gres. These deposits are quite hëtërogënes, conglomëratic with siliceous or kaolinous cement with numerous intersecting stratifications. They are sparsely represented and occur in the north-west of the province in

Kouni and its surroundings. These deposits contain large marine pebbles, generally of small size (2 cm) and formed almost entirely of quartz.

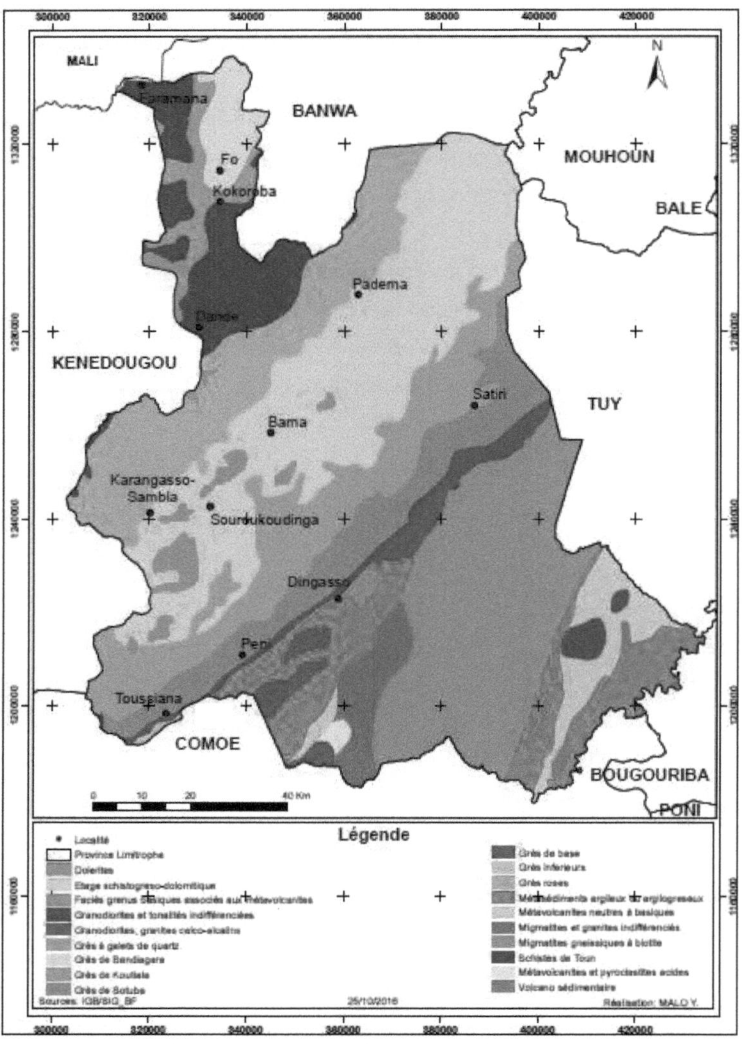

Map 6: Geological formations of the Houet province

2.2. The human framework

The province of Houet is still composed of several indigenous ethnic groups that have gradually settled there. Also, a multitude of other non-native ethnic groups have immigrated to the province. This settlement could be partly due to the physical conditions of the province, more particularly to the morphology of the relief which has always conditioned the settlement of men in a given environment. It is a province with multiple potentialities, favourable to the development of socio-economic activities.

2.2.1. The settlement and population revolution of the Houet province

The province of Houet has been progressively settled by several factors. The history of the settlement of this province, as well as that of most of Burkina Faso, remains unclear.

2.2.1.1. Establishment of populations

It would seem that the oldest settlers were the Bobo-fing from the Xeme century. The Bobo-Dioula would have come much later from Djienne (present Mali) around the XVeme century. The Bobo are different from the Bwa with whom ethnography has long assimilated them. But in no way can the Bobo be linked to the Bwa or vice versa, because the two peoples are absolutely alien in language and customs, but they belong to the same cultural level. After the Bobo came the Toussian in successive waves from the territories between Nielle and Sikasso in the west of Mali, in search of new land and also to escape from the invaders in the 17th century^{eme}. At the end of the XVIIeme century, the Kingdom of Kong was created by the Ouattara and expanded through military expeditions by integrating villages as numerous as those of Bobo, Tiefo, Vigue, Bolon, Komono, Dorobe, Dafin and Bwaba. It was during this invasion that the Bobo-Dioula group was born, according to KONATE S., 1995, which is a crossbreeding between the Dioula and the Bobo. Following these concessions, an allied kingdom was created called Guiriko. Its management was entrusted to the Bobo-Jula who established the Sanou Dynasty.

2.2.1.2. Spatial distribution of ethnicities

These peoples are spread out on both sides of the province. The Bobo group, which occupies a large area in the province, is located in the centre, on the Bobo-Faramana, Bobo-Hounde and Bobo-Dedougou axes. The Senoufo group, composed of the Toussian in the department of Toussiana and the Tiefo in the department of Peni. In addition to these two large groups, there are isolated ethnic groups such as the Vigne in the east in the department of Karangasso-Viguea, the Sambla in the west in the department of Karangasso-Sambla and the village of Bouende and the Dioula in the department of Darsalamy. These indigenous families share this territory with other immigrant ethnic groups, notably the Mossi, the Bissa, the Gourmantche, the Dafin, the Samo...

2.2.1.3.   Socio-political organisation

The social organisation of societies is considered to be hybrid, i.e. it straddles the line between societies with centralised power (e.g. the Mossi) and so-called acephalous societies (the Lobi). However, two types of society without centralized power, namely the village community (Bwa, Bobo, Bobo-Dioula...) and the society with a lineage organization (Ciramba, Curaba, Vigue, etc.) exist in the province. In these societies, the only source of power lies in seniority, which applies to age in the case of an individual and to the authority to occupy the territory in the case of a lineage. The main characteristic of the village community is that the political horizon is limited to the village level; beyond this reality, the people know nothing that can solicit their loyalty. But contrary to what one might think, the chief is not unknown in these societies. He does exist and performs the same functions, with a few exceptions, as chiefs in so-called centralized societies. He represents and directs the community and performs political, economic and socio-religious functions (GUYOT-SAVONNET C., 1986). Apart from these types of organisation, the Dioula have a centralised power.

2.2.2. Socio-economic activities

The economy of the province is mainly based on agro-silvicultural activities, trade and industry.

2.2.2.1.   Agriculture

As in other provinces, agriculture remains the main activity of the province's population, particularly those in rural areas, and occupies a large workforce. Three types of agricultural systems are still practiced in the province. These are intensive, semi-extensive and extensive systems. The main crops are maize, millet, white and red sorghum, fonio, cotton, rice, sesame, soya, etc. It is the anthropic activity that most affects geomorphological forms. This growth is possible thanks to agricultural equipment for an agriculture that is increasingly evolving towards semi-mechanisation. The province has 33% of the region's agricultural equipment according to the INSD, 2006.

2.2.2.2.   Breeding

It occupies a prominent place in the economy of the province after agri-culture. The indigenous farmers are both farmers and herders. Several production systems are observed in the province. According to INSD (2006), Houet accounts for 53.57% of the cattle population and 69.22% of the sheep population in the region. In terms of ruminant breeding (cattle, sheep, goats), there are two types of production systems. The traditional extensive systems (transhumant and sedentary) and the improved systems (semi-intensive and intensive). For example, in 2008 the Commune of Lena ranked second in the country in terms of transhumance of cattle, with 12,300 head according to the Direction Generale des Statistiques Animales. Pig farming is also developed in the province. It is also semi-intensive and intensive around Bobo-Dioulasso and of the traditional or village type, which accounts for 80% of total production. As for poultry farming, it is both traditional and practiced by both men and women and plays a significant socio-economic role. It is also modern but is particularly practiced in the commune of Bobo. In 2008, there were 56 poultry farms in the Hauts-Bassins region, 55 of which were in Bobo alone.

2.2.2.3.   Trade

The province, through the city of Bobo, an international crossroads, has a fairly developed trade. It had 18 markets in 2008. A large number of national and foreign trading houses have their headquarters in Bobo-Dioulasso. The Bobo Chamber of Commerce counted 180 trading establishments in 2008. Their activities are generally focused on import-export, the sale of textiles, construction materials, automobiles and electrical appliances. There are two types of trade: the structured or modern trade that is regularly registered in the trade register and has a licence; it generally operates in wholesale sales and participates in public tenders. In addition, there is the informal trade, which is the most important and is practised in all localities.

Partial conclusion

As there are not many geomorphological maps in Burkina Faso, those that do exist are at small scales. Thirteen gëomorphological units are identified in our study area. The sandstone plateaus are dominant. These geomorphological units are underlain by the sandstone formations of the south-eastern edge of the Taoudeni sedimentary basin. Other units are based on the crystalline basement in the east of the province. The relief of the province of Houet is uneven and the hydrographic network is quite dense with the Mouhoun as the main river. The Comoe River, which rises there, is also important. Several classes of soils exist and are mostly suitable for agriculture. The climate is south Sudanian and the average annual rainfall is over 900 mm. As for the flora, it is mainly composed of wooded savannahs with shrubs and some open forests.

PART TWO

GEOMORPHOLOGY AND ORGANISATION OF THE RIVERS OF THE HOUET PROVINCE

This part deals with field and secondary data and satellite images. It also includes two intrinsically geomorphological chapters. One of the two chapters deals with geoforms and the other with geomorphological units and the organisation of the hydrographic network of the Houet province.

Chapter 3: Typology and mapping of the landforms of the sandy plateau of the Houet province

It is about the different irregularities of the terrain and their characteristics: altitude, lineaments, contour lines...

3.1. The development of the sedimentary formations of the Taoudeni Basin

The Sedimentary Basin was gradually built up over a long period of time with occasional tectonic movements.

3.1.1. Presentation of the Taoudeni Sedimentary Basin

The establishment of the Taoudeni sedimentary basin, according to KOUSSOUBE Y. (2013), is part of a dynamic that started around 1000 Ma in the Neoproterozoic and Paleozoic and ended in the Carboniferous. During this period, a number of sedimentary basins including Taoudeni, Tindouf, Volta, Senegalo-Mauritanian, Bove, Ghana, Reggane, Tamesma, Almet, Oued Mya, Illizi and Murkzuk, were developed on the West African craton. Subsequently, other detrital or fluvio-lacustrine basins appeared in the Tertiary (Gondo, Nara, lullemmenden, Chari-Baguirmi...) and formed deposits of variable thickness in discordance with the earlier formations. The Taoudeni basin is one of the largest Precambrian and Neoproterozoic basins in Africa and covers about two million km^2 . It covers a large part of Mali, Mauritania, the two Guineas and extends into Burkina Faso, Algeria, Senegal and Sierra Leone (OUEDRAOGO C., 1983). The south-eastern edge is located on the right bank of the Niger River, which crosses the basin over a length of nearly 1500 km along a west-east axis. The surface area of the south-eastern edge of the Taoudeni sedimentary basin (Mali-Burkina Faso) is about 260,000 km^2 of which 45,000 km^2 are located on Burkina Faso territory, i.e. less than 20%. It is partially covered by Tertiary and Quaternary surface formations (DEROUANE J., 2006). This basin represents 74% of the surface area of the Houet province according to the geological map of HOTTIN G. and OUEDRAOGO O. F. (1975). The remaining % is occupied by crystalline formations in the south-eastern and eastern parts of the province.

3.1.2. The lithological succession of the south-eastern edge of the Taoudeni Sedimentary Basin

Like the other sedimentary basins, it has a lithostratigraphy that varies considerably depending on whether it is in the northern, western, southern, central or eastern part. In Burkina Faso, these formations have been divided into four main groups: Banfora, Falaises, Bobo and Bandiagara which is not développë in this work.

3.1.2.1.   The Banfora group

It is at the base of the stratigraphic sequence and comprises a single thin, limitedly extended dëtritic formation that corresponds to a marine sëdimentation. In the Banfora area, it forms a narrow band that extends in the form of a tongue for about 30 km inside the crystalline basement where it forms a faulted synclinal structure. The thickness of the formation is estimated at 300 m. The Gres Inferieurs formation represents a residual evidence of a shallow marine sedimentation having undergone tectonic events prior to the deposition of the overlying formation (OUEDRAOGO C., 2006).

3.1.2.2.   The cliff group

This group comprises two essentially greseous formations of very variable granulometry, very coarse, conglomerated to very fine. On the whole, it is a shallow marine sedimentation of epicontinental type, with more or less marked fluvial episodes. It is subdivided into two formations according to OUEDRAOGO C. (2002 and 2006) as follows

> **the formation of the Gres Kawara-Sindou (GKS)**
It constitutes the essential part of the gresous cliff at the edge of the basement. In the Banfora region, it rests unconformably on the Lower Gres. Elsewhere, it rests directly on the bedrock; very thick in the western region of Negueni-Kawara (about 350 m), it thins at the base towards the east in the Toussiana region (about 60 m), then disappears north-east of Bereba. Further north, to the east of Dedougou, it appears again as a small outcrop near the village of Fie;

> **the formation of the Takaledougou Glauconitic Fine Gres**
It forms the top of the cliff and rests unconformably on the underlying Kawara-Sindou Gres formation; towards the north-east, it directly covers the crystalline basement. With a maximum thickness of 600 m, it begins with fine or very fine, silty, often glauconitic, grey, schistose flow and desiccation figures alternating with coarse, microconglomerated, often glauconitic, grey, platelet flow gres-quartzites. Above this comes a level of greenish argillite with a schistose flow, up to 80 m thick, fine quartzite sandstones and siltstones, generally glauconitic and micacës, well bedded with very rare oblique tabular stratifications and polygonal desiccation figures, about 80 m thick (OUEDRAOGO C., 2002). The upper part, which can reach 400 m, presents a remarkable ruiniform aspect and is often largely gullied by the upper formation. Subsequently, a new marine transgression took place with, at the beginning of the sedimentation, periods of emersion. This was followed by the establishment of relatively deep platform marine conditions without continental influence, with relatively weak current regimes. The upper part corresponds to an increasingly shallow environment dominated mainly by wave action (OUEDRAOGO C., 2006).

3.1.2.3.   The Bobo Group

This group, which includes the Houet province, has been subdivided into five subgroups according to OUEDRAOGO C. (1983 and 2006) whose work has been widely used by KOUSSOUBE Y. (2013). Thus, both have stratigraphed and described the granulometry of these groups.

> **The formation of quartz granular sandstone**

It begins the Bobo-Dioulasso Group formations and quite clearly ravines the underlying ones. This formation is thought to be formed by fluvial deposits in braids in which the Aeolian influence is important. It outcrops quite widely to the west of Bobo-Dioulasso. Its thickness varies between 0 and 500 m approximately. Its base is not very well known; most of the formation is made up of fine to medium feldspathic gres-quartzite with microconglomerates. Locally, there are intercalations of very fine, well-graded, finely bedded and micaceous sandstone and rare, well bedded reddish claystones. They contain small millimetre-sized pebbles of rolled quartz and rare oblong decimetre-sized pebbles of quartz-grey. Ripple-marks and more generally oblique, tabular, decimetrical to metric laminations with flat sheets and metric-sized gutters are encountered. The sheets of the laminations are often underlined by quartz granules.

Silt, Clay and Carbonate formations

This formation, which is poorly exposed, is in agreement with the underlying quartz granular sandstone formation. It begins with about ten metres of coarse reddish-brown, glauconitic sandstone, containing thin intercalations of very fine, silty, pink, well bedded and micaceous sandstone and greenish argillite. This is followed by an alternation of fine feldspathic and glauconitic, ash-grey to pink, reddish siltstone with a plate-like flow and greenish, micaceous, schistose-flowing argillite.

This clay-siltstone alternation seems to constitute the main part of the formation with in places large beds of dolomite and dolomitic limestone with stromatolites. The carbonate horizons have so far been considered as lenticular levels. The hydraulic drillings carried out in the framework of various village water supply projects have intersected carbonate levels in several places, but their later continuity could not be really well established. On the other hand, these boreholes have revealed at least four stratigraphic levels of carbonate horizons. These carbonate rocks present two facies: granular carbonate rocks (with oncoliths, endoclasts, rare ooliths), homogenous carbonate rocks which include stromatoliths, rare terrigenous detrital elements, brecciated passes and passes with wedge-shaped drying cracks. The total thickness of the formation may exceed 300m. The transition from fluvial to free marine sediments appears to be gradual and corresponds to a new marine transgression. The sedimentary and organic structures indicate a shallow, subtidal to intertidal marine sedimentation environment with warm waters of variable salinity, low agitation and low detrital inputs.

> **The formation of the fine pink Gres or Bonvale Gres**

It was established after the first two formations and can be followed from west of Bobo-

Dioulasso, to the north-east in the Nouna area. It is concordant with the underlying formation. It is about one hundred metres thick and is essentially made up of fine to very fine, micaceous and glauconitic gres-quartzite, finely bedded, with numerous symmetrical ripple-marks with undulating or linguloid crevices. Clay pellets and load casts can be seen locally. This formation represents a period of shallow marine sedimentation, entirely terrigenous.

All of the above formations are injected with dolerites in the form of sills or dykes. The veins are not very powerful (20 m) and it is sometimes possible to follow them for tens of kilometres. Their directions are very varied. In the landscape, the dolerite intrusions are remarkable by their high altitude.

> **The Samandeni-Kiebani Siltstones, Claystones and Carbonates**

It is a formation that outcrops very little. It corresponds to a vast area pёпёplaпёe covered by clayey alluvium and latёrites. The rare outcrops appear in the Samandёni area, on the sides of small isotee hills, protectedёgёs from erosion by a latёritic or doteritic cover. It lies in concordance with the underlying pink fine sandstone formation. The total thickness of the formation is estimated at 450 m. Once again there is a frank marine sёdimentation in a low ядПё environment with low dёtritic inputs.

> **The formation of the Siltites and Gres-quartzites of the Fo Pass**

This formation outcrops well at the foot of the Fo cliff and appears to rest concordantly on the preёcёdent formation. Its top is raptured by the overlying sёrie; it would therefore constitute the terminal formation of the Bobo Group. Its thickness could reach 50 m. This formation provides a transition to an even shallower marine environment with dёtritic terrigenous sёdimentation and pёriods of dёp6ts that mark a tendency to immersion.

Figure 5: The stratigraphic column of the south-eastern Taoudeni Basin

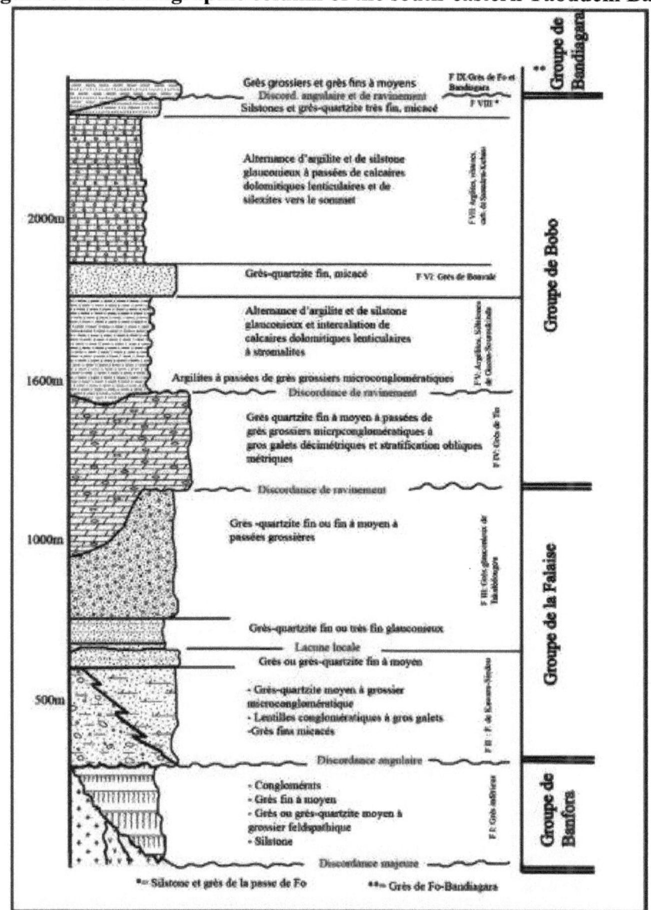

Source: OUEDRAOGO C., (2006)

3.1.2.4.   The dips that affected the sedimentary layers

According to SANOU D. C. (2008), in the centre of these sedimentary basins, the structures are aclinal or horizontal with zero dip. However, there are also monoclinal sedimentary structures where the layers dip in a single direction. In the Houet province, several dips of different orientations are observed between the layers. In the same locality they differ from one level to another. The geological sections made on the Banfora and Bobo sheets by OUEDRAOGO C. (1983) allow to appreciate these dips.

Amongst these sections, the Toussiana section, oriented north-west/south-east, was raised to the east of this village located on the Bobo-Banfora axis. The sedimentary layers are affected by dips of 6° north-west in levels 1, 2, 3, and ; 4° in levels 5, 6, 7 and 9 and north-west.

43

Figure 6: The geological section of Toussiana

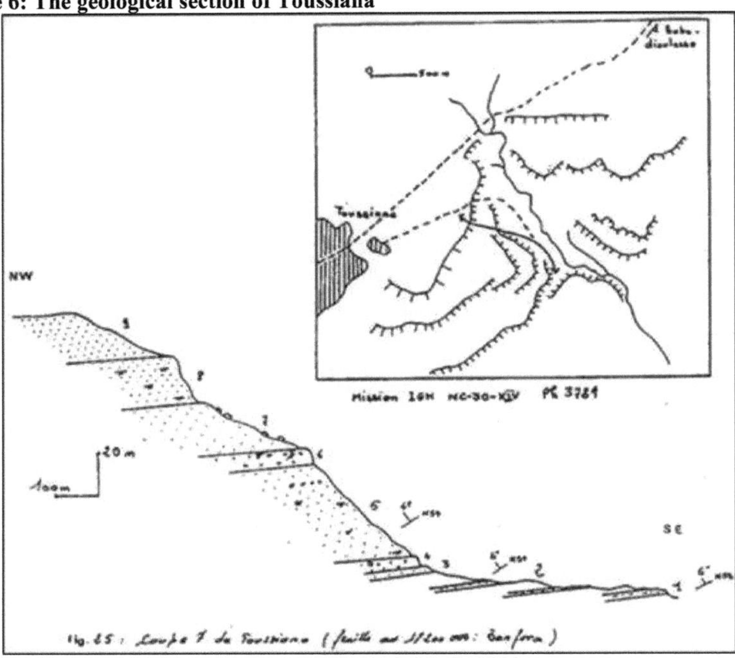

Source: OUEDRAOGO C., 1983

Table 2: The explanatory note of the Toussiana geological section

Level	Description	Level	Description
1	clay-gravel level	6	coarse conglomerated sandstone
2	sporadic outcrop of fine purplish red sandstone with pink or red fracture	7	soft slope mask
3	fine to medium gres-quartzite	8	very fine to medium gres-quartzite
4	medium to very coarse sandstone-quartzite	9	very fine pink or red gres
5	fine to medium gres-quartzite, grey-black	-	-

The Tiara section of the Guena-Souroukoudinga Clay, Siltstone and Carbonate, west of Bobo-Dioulasso, dips 11° south-southwest and includes levels 1, 2, and 3. However, level 10 has a dip of 10° and is oriented south-southwest. This section is located about 2 km east of the village of Tiara, 30 km from Bobo-Dioulasso on the Bobo-Orodara road, in a dolomitic limestone quarry.

Figure 7: Geological section of Tiara

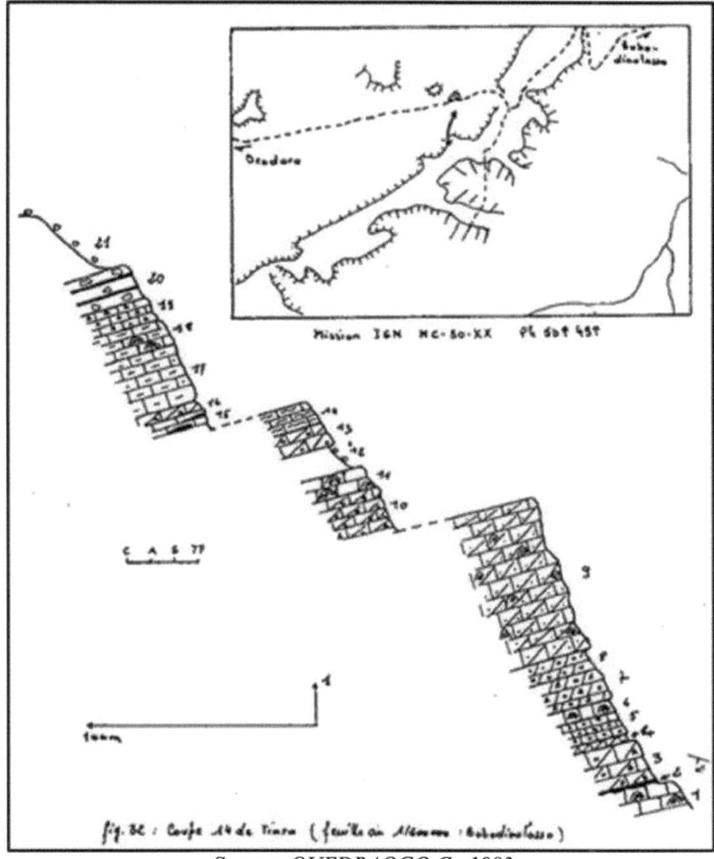

Source: OUEDRAOGO C., 1983

Table 3: The explanatory note of the Tiara geological section

Level	Description	Level	Description
1	white cryptocrystalline limestone	12	limestone scree
2	thin level of sandy limestone with fine, slightly undulating horizontal laminations	13	grey dolomitic limestone with oncoliths
3	massive dolomitic limestone bench with purple breccia	14	beige limestone with a very slightly undulating horizontal laminar structure with limestone flow
4	light grey cryptocrystalline dolomitic limestone	15	beige grey limestone with thin lenticular black flint
5	light grey limestone well lite	16	thin level of yellow brecciated dolomitic limestone
6	white cryptocrystalline limestone	17	well bedded greenish clayey limestone, with sparse flows
7	brecciated dolomite with oncoliths	18	microcrystalline clayey limestone, grey-black
8	ootite grey dolomite	19	Brechian grey limestone
9	whitish to mauve silty dolomitic limestone well bedded	20	siliceous rock, whitish or greenish
10	crystalline dolomitic limestone	21	laterite scree
11	whitish dolomite with small bushy stromatolites	-	-

3.2. The topography of the sandstone formations

The database of the Institut Gëographique du Burkina (IGB) at 1:200,000 in combination with the digital terrain model (DTM) allowed the production of maps of heights, contour lines, slopes, Triangular Irregular Network (TIN) and lineaments.

3.2.1. Topographical inequalities

They refer to the different elevation variations that exist in the geological formations that explain the geomorphology. The elevation map below shows the differences in elevation from one area to another. The map was produced by classifying the SRTM image with ArcGis 10.0 software.

The lowest altitudes are found near the village of Segueri to the north of Bama and continue north-eastwards along a corridor to the latitudes of villages such as Kimini, Koledougou and Soma, all in the commune of Padema, which remains the lowest in the province of Houet. In the opposite direction, the high altitudes are found in the south-east in the villages of Tien and Tapioko in the communes of P'ni and Toussiana respectively. According to the results of the SRTM image processing, the altitudes vary between 256 and 694 m. On map 7 below, the highest altitudes are shown in red and the lowest in dark blue. A comparison with the points on the IGB that vary between 272 and 668 m shows that this height map has some accuracy.

The average altitude calculated by averaging the five lowest altitudes plus the five highest altitudes divided by two gives 452 m. This is [(256+275+272+273+283/5) + (694+668+662+568+571/5)]/2= 452 m.

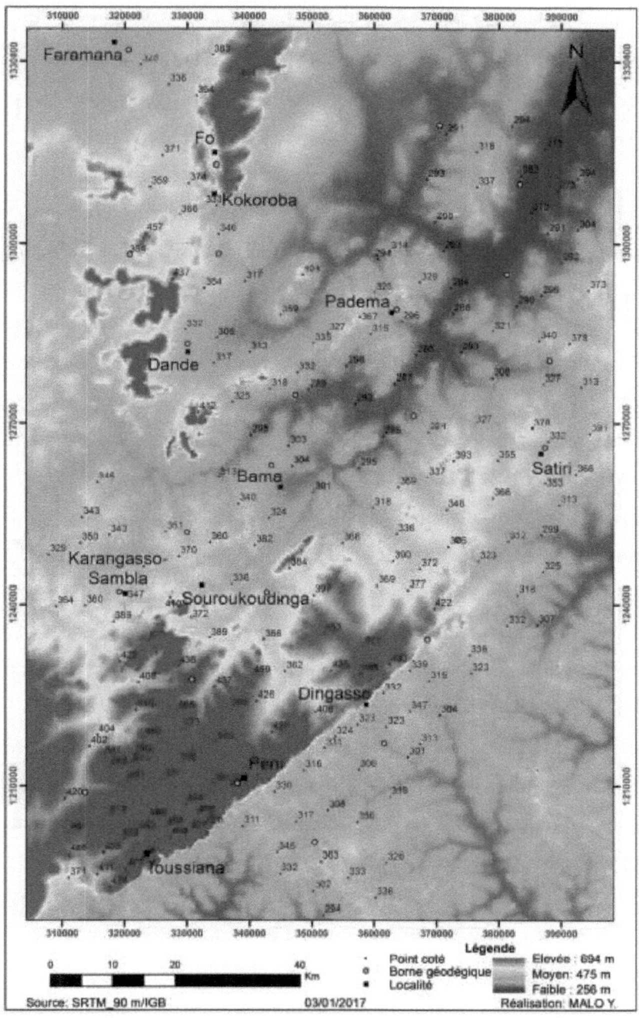

Map 7: Altitudes in the south-eastern edge of the Taoudeni sedimentary basin

3.2.2. The contour lines

Contour lines or isolines are lines that connect all points of equal altitude on a given terrain. Their realisation required the use of a raster image (SRTM). The ArcScene extension of ArcGis 10.0 allowed for a 3D analysis of the raster data. The ArcToolbox spatial analysis tool, especially the Surface extension, automatically generated contour lines as shape files. These contours are equidistant of 10 m. The results obtained show that the relief is quite uneven. The south-western part of the area seems to be the most rugged with prominent features such as the Trikma and Tiara mountains. A little to the south, a long chain of high reliefs, less wide, originates southwest of Peni. This is the Banfora cliff. It extends south-eastwards through the village of Dingasso and begins to taper off at Borodougou, east of Bobo-Dioulasso, before disappearing completely west of Kotedougou, near the village of Dafinso. In the extreme west, near the village of Diofola in the commune of Karangasso Sambla, there are also high reliefs, but less extensive than in the south and south-west, which extend towards Lanfieracoura, in the commune of Dande, while covering the west of the village of Natema, in the commune of Bama. The Kantolo and Tiara mountains are located in this area. Further north of Dande, in the commune of Fo, there are some high reliefs that extend to the border with Mali: the Bandiagara Plateau, which we observed in the field. Others are however isolated in the commune of Faramana.

The bas-reliefs concern part of the northern part of the basin as well as the south. They correspond to the plains and extend from the commune of Bama to that of Padema in the north-east. Another plain begins in the commune of Dande, covers the west of that of Fo, continues in that of Faramana and continues into Malian territory.

Thus, two major types of landforms share the province. The plains in the northern, north-eastern, southern and south-eastern part and the plateaus in the southern, south-eastern and south-western part. Field checks have enabled us to label certain eminences such as Sadinakourou, Samandenikourou, the Kouari and Tolo "mountains".

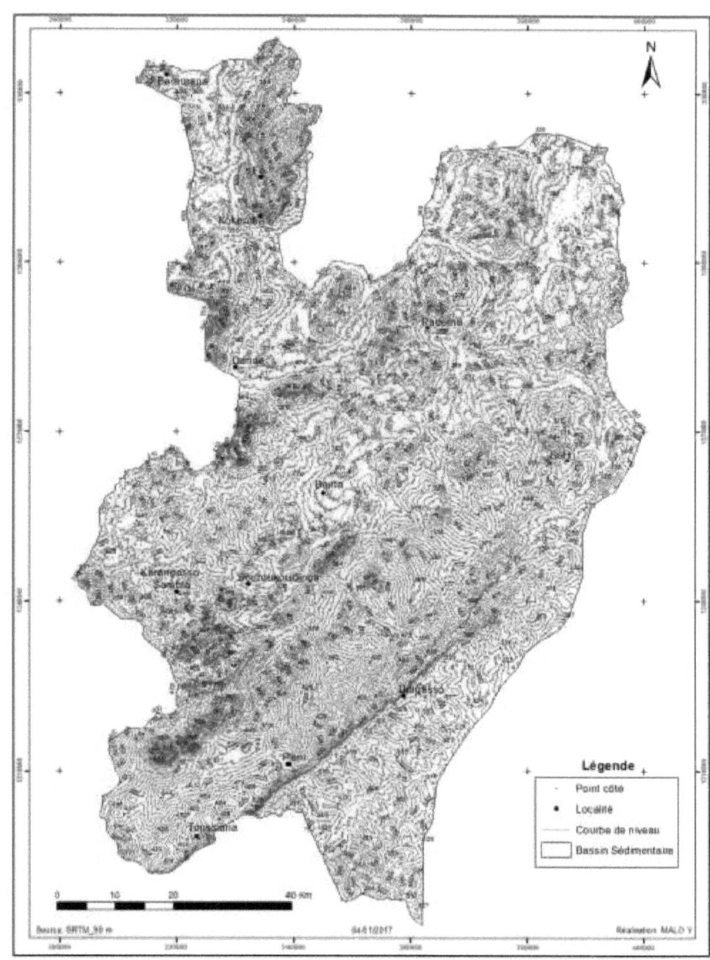

Map 8: Topography of the commune of Houet

3.2.3. The slopes on the south-eastern edge of the Taoudeni sedimentary basin

These basins are conducive to the formation of three types of landforms: plateaus, cuestas and plains. These reliefs are presented differently depending on whether they are in the centre or on the edge of the sedimentary basin. The reliefs, which are created by several factors, have slopes that are sometimes steep, moderately steep or gentle.

The three-dimensional (3D) representation of the slopes made it possible to assess their orientation and degree of inclination. The slope map was created using the ArcScene extension of ArcGis 10.0. The classification was done with the spatial analysis tool of ArcToolbox. The so-called Natural Statistics (Jenks) classification was chosen because it produces the slope (degrade or maximum rate of change of the z-value) of each cell in a raster surface. The results show that three types of eminences are present in the Houet province: cuestas, plains and plateaus. To these ensembles can be added all the evidence of lateritic cuirasses (relics of bauxitic and ferruginous cuirasses, high glacis, medium glacis, low glacis, neoformational glacis) on bedrock or on sedimentary formations. OUEDRAOGO C. (2006) finds that these cuirasses are in place from the quaternary to the present. The slopes vary from 0 to 37°.

In the centre there is an isolated plateau with average slopes of between 10 and 14°. In its western part, gentle slopes (less than 2°) form the transition to a depression. To the south, near Peni to Borodougou, steep slopes with values of over 15° rise up. This is a wall that rests directly on the basic sandstone formations east of Toussiana and Peni. But to the east of Bobo-Dioulasso it rests on the base. It slopes towards the west and the cuesta reverse, which is quite visible on the map, descends into the depression that separates it from the sandstone plateaus to the southwest. To the east, a slope joins the cuesta to the plain below, where the sugar cane plantations of Takaledougou 1 and 2 are located. From Peni to Banfora via Toussiana, an escarpment separates the plain from the cuesta. To the south-west is the highland area, as evidenced by the mostly steep slopes. In the north, in the Fo region, cuestas have the same characteristics as those found in the southern part. However, they are less important than the latter and are strongly dissected according to our field verifications with a cuesta reverse sloping towards the west. In the north-west, near the villages of Dande, Kokoroba, other plateau areas can be observed. They

They all have gently sloping hillsides. To the west, near the localities of Natema, Diofoloma, a chain of plateaus dominates the landscape with hills in places. Two eminences, Sadinakourou and Samandinikourou, dominate the landscape (Photo 1-A and B).

<u>Photo 1 A and B: The cliff of Banfora in its northern and southern part</u>

Photo A: The northern extension of the Banfora cliff with a receding cuesta front due to differential erosion.

Cliché : MALO Y.

Photo B: It is possible to see in light green the plain prepared for sugar cane and the one on which is establishes the village of Takaledougou 2.

The plains occupy the north-eastern, eastern, south-eastern and southernmost parts of the province and are dominated by armoured hillocks and plateaus of lesser importance at certain levels.

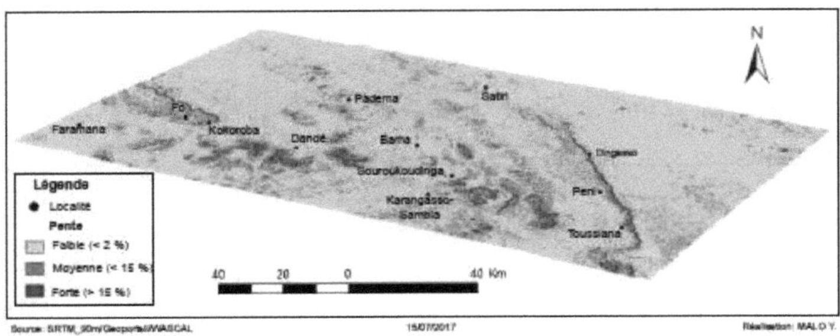

Map 9: Slopes in the Houet province

3.2.4. The different fractures that have affected the Grese plateau

The Taoudeni Basin has been affected by recent to current tectonics in contrast to the apparent calm of the West African craton today. We note with DEROUAN J. (2006) and KOUSSOUBE Y. (2013), that the Burkinabe side of the Taoudeni basin is marked by brittle to soft tectonics. This is at the origin of faults, diaclases and "cliffs".

Several village water supply programmes in Burkina Faso have provided an understanding of the deformation of sedimentary deposits. The conditions of outcrop of these neoproterozoic formations have made it possible, on the basis of geological sections, to highlight faults of regional importance. Thus, in the south-eastern edge of the Taoudeni basin, the following faults were identified

-    a major fault of north-east orientation following the course of the Mouhoun (rising branch) but which gradually straightens out until the confluence with the Sourou (Lery);

-    a north-east fault becoming north-south like the first one;

-    A north-south mëridian fault running from Banfora up into Mali. Along this fault and dolerite intrusions, numerous ëmergences (springs) and artësian drillings have been identified (KOUSSOUBE Y., 2013).

The lineaments map below shows these fractures. For its realisation, Landsat 8 bands 7-6-5 were used to perform a Principal Component Analysis (PCA) or Hotelling transform, which is a technique to enhance a multispectral image for geological interpretation (BIEMI J. et al., 1991). The choice of these three bands is explained by their spectral signature characteristics. Indeed, it is a question of creating new channels (**neo channels**) summarising the information contained in several bands. In short, it is a process aimed at statistically maximising the amount of information or variance of the original data in a limited number of components (CHEREL J.-P., 2010). The PCA is performed with ENVI4.5. It was then imported on PCI Geomatica V9.1 for an automatic extraction of lineaments. The principle consists in establishing lineament maps by systematically noting all the linear and circular structures visible on the image (OUEDRAOGO C., 1983). These shapes can also be the boundaries of plots, or the roofs of houses and huts. When superimposed on the ACP image, these lines do not correspond to the roofs of houses, roads or the hydrographic network. However, the lineaments are automatically sorted from the other assimilated lines. The image below shows the lineaments in red and the other lines in white.

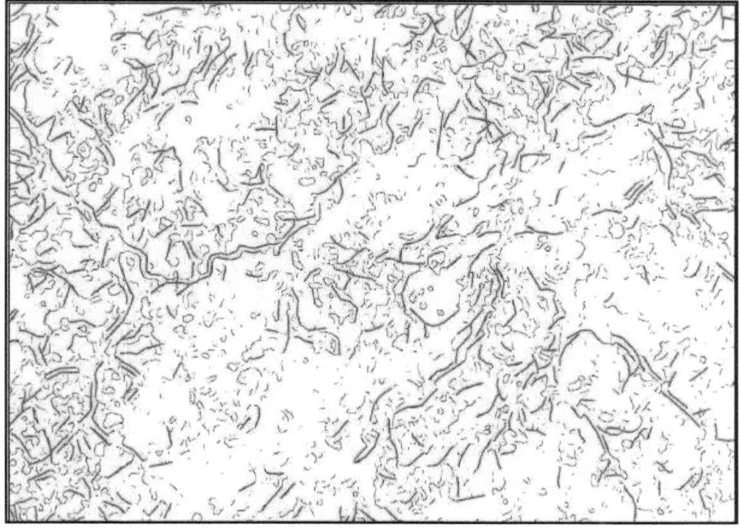

Map 10: Raw lineament extraction

This confusion has been overcome by cross-referencing the lineament map with the 1:200,000 LGB road and river network map. These lineaments have the advantage of covering all the sedimentary formations of the Houet province, as almost all the lineaments in the province are localised and have a specific purpose.

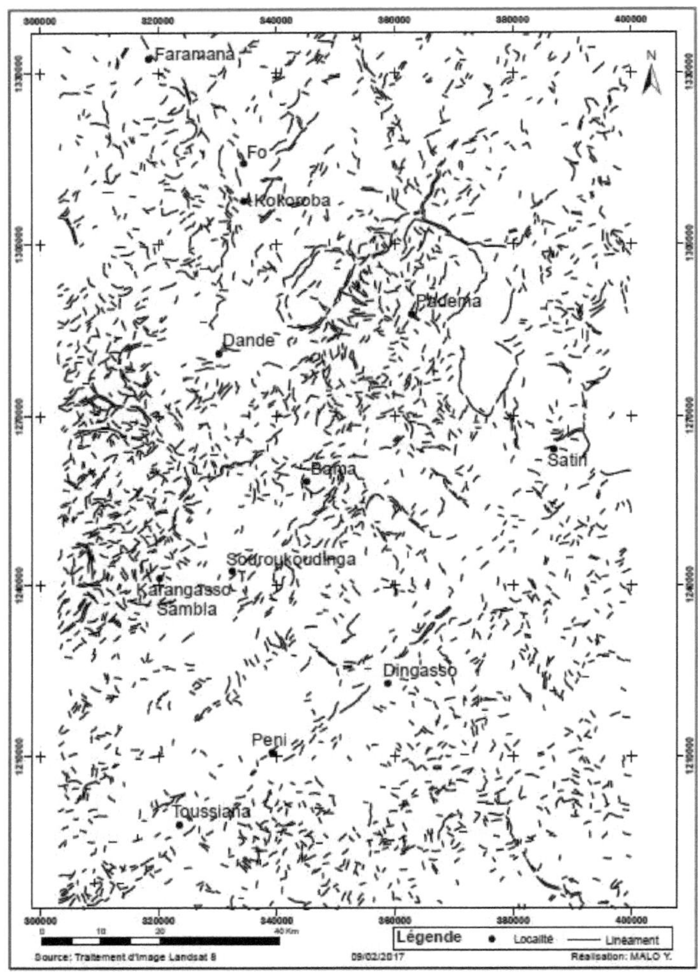

Map 11: The lineaments in the sandy plateau of the Houet province

The lineament density map shows that in the plateau and high relief areas, lineaments are less important but on the plains they are quite evident. Indeed, the Burkinabe part of the Taoudeni sedimentary basin has been affected by two major tectono-orogenic events (KOUSSOUBE Y., 2013) explaining these fractures. These are :

- the Pan-African orogeny (600 Ma) which reactivates the main basement faults; NE to NNE to EW movements affect the basement (Burkina Faso) and the interior of the Taoudeni basin (Nara Trough).

- The opening of the Atlantic in the Lower Jurassic, a period during which dolerite intrusions, which began in the Terminal Paleoproterozoic (1810 Ma- Mesoproterozoic), reached their peak in intensity throughout West Africa.

Map 12 below shows the density of lineaments in the province. Additional processing has allowed the density map to be superimposed on the ACP image. On the plateaus and plains in the south they are less dense, but the plains and to a lesser extent the plateaus in the northern part have fairly dense lineaments.

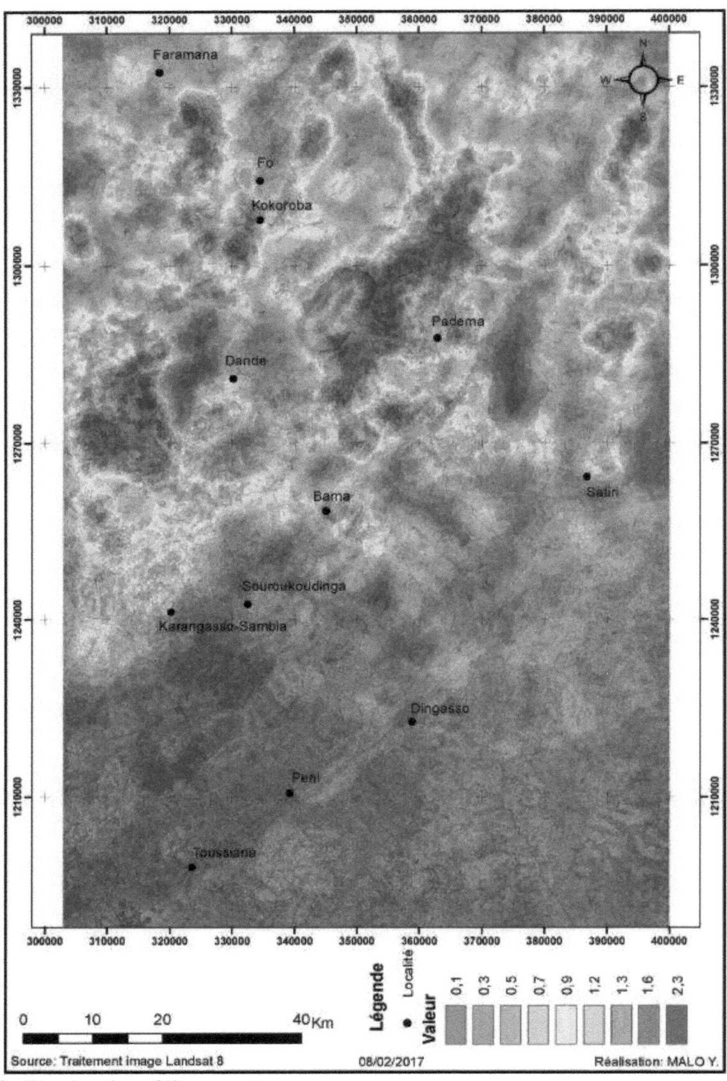

Map 12: The density of lineaments

3.2.5. The fracture rose diagram

It was designed to characterise the direction of the different lineaments in the sandstone formations of the Houet province. It is a semicircular diagram made from spatial data in (x ; y). It allows us to appreciate the orientation and spatial organisation of the lineaments. These coordinates are obtained in ArcGis 10.0 with the Split extension which allows to split the different lineament segments. The file is exported to CAD in DXF_R2010 format. This was then imported into RockWorks15 geology software for the construction of the directional rosette. The rosette was constructed as a percentage of the number (frequency) of lineaments with the directions arranged in 15° classes (see Figure 6).

Figure 8: The directional rosette of the lineaments

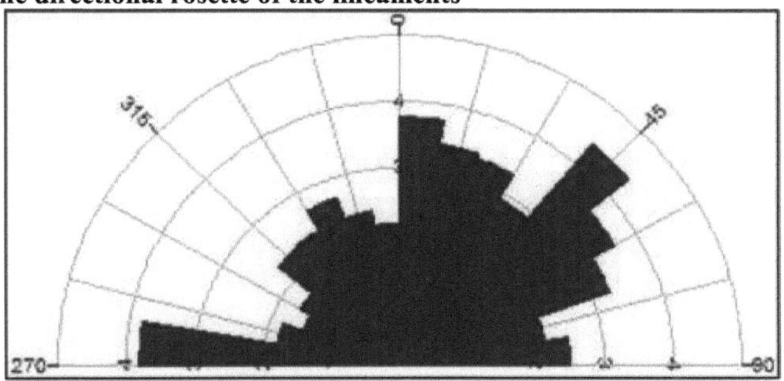

Source: Landsat 8 image processing

Analysis of the diagram shows that the lineaments run in several directions. These are the northern directions (N0°-N10; N10°-N20°; N20°-N30°; N30°-N40°; N40°-N-E50°). The N-E direction which corresponds to 45° is the most representative. In addition other directions (N50°-N60°; N60; N70°; N70°; N80°) are quite important. Going west, there are the directions W270°-W280°; W280°-W290°; W290°-W300° W300°-W310°; W310°- W-N320° which is the W-N direction (315°). The last directions are West-North (W-N320°-W-N330°; W-N330°-W-N340°; W-N340°-W-N350°; W-N350°-N0°).

In general, four observed directions are less important: N70°; N80°; N80°; E90° and those W280°-W290°; W290°-W300°. The N45° direction is the most abundant in the sandy plateau. OUEDRAOGO C. (1983), had highlighted this direction but in a more restricted framework, i.e. at the scale of the Guena zone, 40 km west of Bobo-Dioulasso. In addition to this direction, those W270°-W280° and N0°- N10; N10°-N20°; N50°-N60° are quite important.

3.2.6. The 3D relief map

This is a relief map rëalisëe thanks to the ArcScene extension of ArcGis 10.0. It is a svnthese map that presents the relief of the province in 3D. It is prëcisëment of the Triangular Irregular Network (TIN). The TIN is a vector representation of surfaces generated by a triangulation method that can be easily extracted from any spatial data (ILBOUDO P. S., 2010). It allows to answer most of the applications because of its respect in the precision of the data. It is suitable for 3D display of topographic data. The TIN realized under ArcScene favours a simulation of the real terrain. But in this case, there was a slight exaggeration of the heights.

The maximum difference allowed by default between the height of the input raster and the height of the output TIN is 1/10 of the Z range of the raster. We have set this to 1.5/10 because 10/10 was too much. The maximum size of a TIN that can be used is between 1.5 and 20 million nodes. The nodes are TIN points whose densification allows this representation which tends towards reality. Here we have used 20 million nodes. For example, cuestas and cuesta reverses are very well visible as well as escarpments. The plateaus, buttes and foothills can also be identified in the south-east and east (see map 13). Thus, the plains, plateaus, cuestas, cuesta reverses, witness buttes and fore buttes can be appreciated. The south-western, central and northern parts are occupied by plateaus, while the south-eastern, north-eastern and southernmost parts are areas of plains dominated by medium-sized hills of more than 350 m in height and foothills.

Source: Image processing SRTM_90 m

Map 13: The ground truth simulation of the study area

Chapter 4: The geomorphology of the Grese plateau

In this chapter, the aim is to highlight the geomorphological units of the sandstone formations in the Houet province. Landsat 8 images were used and field verifications were used to confirm or deny the training plots.

4.1. Acquisition, description and processing of Landsat 8 images

These are Landsat8 images whose colour composition of the bands allows the different elements of the physical environment to be highlighted.

4.1.1. The choice and acquisition of Landsat 8 images

The choice of Landsat 8 images is due to their availability and free of charge. Landsat 8 images have been chosen because of their availability and their free nature. The multitude of channels in Landsat 8 images provides sufficient geographic information. Each spectral band of the Landsat images appears in shades of grey and corresponds to a portion of the electromagnetic spectrum. Three scenes cover the study area according to the Landsat image assembly table, namely 197/52, 196/52 and 196/53. These scenes were obtained from the Division du Developpement des Competences de l'Information et du Monitoring de l'Environnement (DCIME) which is a section of the Secretariat Permanent pour la Conservation de la Nature, de l'Environnement et du Developpement Durable (SP/CONEDD). The three scenes, all from 2014, are detailed in Table 2 below.

Table 4: Landsat 8 scene code decryption

Scenes	Type of image	Path	Row (Rangee)	Year	Month	Date download
lc81960522014070	Landsat 8	196	52	2014	02-March	2014-03-11
lc81960532014070	Landsat 8	196	53	2014	02-March	2014-03-11
lc81970522014093	Landsat 8	197	52	2014	25 March	2014-04-03

The choice of scenes for the March 2014 period is explained by physical factors. This is a province with acceptable vegetation cover and rainfall. Beyond or below this period, the quality of the images could be affected either by nebulosity for rainy seasons that set in quite early in the province or by the vegetation cover that, until after the rainy season, still covers the bedrock. This argument is also reinforced by the need for field verification which is more optimal during this period of the year.

4.1.2. Spectral characteristics of the bands

According to the Quebec Landsat 8 (2015) satellite image mosaic interpretation guide, they are produced from the following bands and filters:

- band 6: shortwave infrared applied to the red filter ;
- band 5: near infrared applied to the green filter;
- band 4: red applied to blue filter;

The visible and near-infrared spectral bands are widely used in optical remote sensing, as they each provide different and complementary information:

• Band 5 (mid-infrared, 1.55 pm to 1.75 pm) of Landsat-5 and Landsat-7 images corresponds to Band 6 (shortwave infrared, 1.57 pm to 1.65 pm) of Landsat-8 images. It is sensitive to soil and vegetation moisture, detects chlorophyll, has high contrast and is insensitive to atmospheric effects;

• Band 4 (near-infrared, 0.76 pm to 0.90 pm) for Landsat-5 and Landsat-7 become Band 5 of Landsat 8 (near-infrared, 0.85 pm to 0.88 pm), which is sensitive to the structure of vegetation cover.

• Band 3 (red, 0.63 pm to 0.69 pm) for Landsat-5 and Landsat-7 are Band 4 (red, 0.64 pm to 0.67 pm) for Landsat-8. It provides clear distinctions between vegetated and non-vegetated areas. With some penetration through water, shades attributable to suspended particles or shallow areas can be seen. Table 3 below shows the Landsat 8 bands with their spectral range and spatial resolution.

Table 5: Landsat 8 bands

BANDS	Colour	Wavelength in gm	Spatial resolution in metres
Band 1	Coastal aerosol	0.43 - 0.45	30
Band 2	Blue	0.45 - 0.51	30
Band 3	Green	0.53 - 0.59	30
Band 4	Red	0.64 - 0.67	30
Band 5	Near infrared	0.85 - 0.88	30
Band 6	SWIR 1*	1.57 - 1.65	30
Band 7	SWIR 2	2.11 - 2.29	30
Band 8	Panchromatic	0.50 - 0.68	15
Band 9	Cirrus	1.36 - 1.38	100 * (30)
Band10	Thermal infrared 1	10.60 - 11.19	100 * (30)
Band11	Thermal infrared 2	11.50 - 12.51	100 * (30)

*Short Wave Infrared

The knowledge and understanding of these bands is important because it offers a possibility to combine them with the spectral signatures of rocks and minerals. METELKA V., 2011 has ëlaborë a stratigraphic library of rocks and minerals in western Burkina Faso with their spectral correspondence. His work shows that the reflectance of rocks is proportional to the increase of their SiO_2 content and inversely proportional to their mafic mineral content. Table 4 presents some of these results.

Table 6: The spectral signature of rocks and their constituent elements

Minerals/rocks	Wavelength in gm
Soil, iron shell, vegetation	0,3-2,5
iron	1,0
Fe and Mg-OH that make up chlorites and amphiboles	2.25 and 2.32
clay	2,2
the group of granites with Al-OH absorption	2,2
phyllo-silicates kaolinite/ smectite	2,2

volcano-sedimentary rocks showing an absorption band in Al-OH	2,2-2,17
fresh surfaces of the Taoudeni gres showing a kaolinite absorption band	2,2-2,17
the reddish to yellow colourations of the gres are related to the presence of hematite and goethite in the matrix, which is documented by the absorption bands of ferric iron	< 0,6-0,65

Source: METELKA V. (2011)

4.2. Landsat 8 image processing

These are the various image processing steps that include enhancement and filtering.

4.2.1. Image pre-processing

The preliminary phase of processing satellite images consists of eliminating radiometric noise in the bands and correcting geometric distortions to make them perfectly superimposable on existing maps. For our images, no geometric correction was necessary; the acquired image was ortho-corrected. Indeed, these images were already used by the SP/CONEDD, so any imperfections that might exist were eliminated. The choice of the Mars period allowed to ignore possible atmospheric disturbances.

4.2.2. Image mosaic

This ëtape consisted of using all possible grey level values (i.e. from 0 to 255 for an 8-bit image); it is performed from histograms of the distribution of the image pixels according to the grey level values for each band. An image with pixels distributed over a range of 100 grey levels will have much less contrast than an image with pixels over a range of 255 grey levels, which does not allow for optimal differentiation of the ëléments in the image. The three scans were initially combined to give an image. The pixels were harmonised by fixing the scene lc81970522014093 which covers the largest portion of the study area and has an acceptable grey level. The bands of the same number on each of the three scenes are mosaicked together (1-1-1; 2-2-2; 3-3-3...).

4.2.3. The processing of the image itself

This is the stage where we identify the different geomorphological units with ENVI.

4.2.3.1.   Colour composition

It consists in combining three bands in the three channels: red, green and blue (RGB). The combination makes it possible to treat several themes and to highlight as many elements of the physical environment. For this geomorphological study, the composition 7, 5, 3 is made in the red, green and blue channels respectively. It corresponds to the 7, 4, 2 composition of Landsat 7. In fact, Landsat 8 bands 5 and 3 now reflect in the near infrared and in the visible range. Band 7 still reflects in the short-wave infrared. According to KABORE D. (2015), the colour composition of bands 7, 4 and 2 became 7, 5 and 3 with Landsat 8 highlights lithology, armour outcrops, contact zones between different rock formations, plant species alignments.

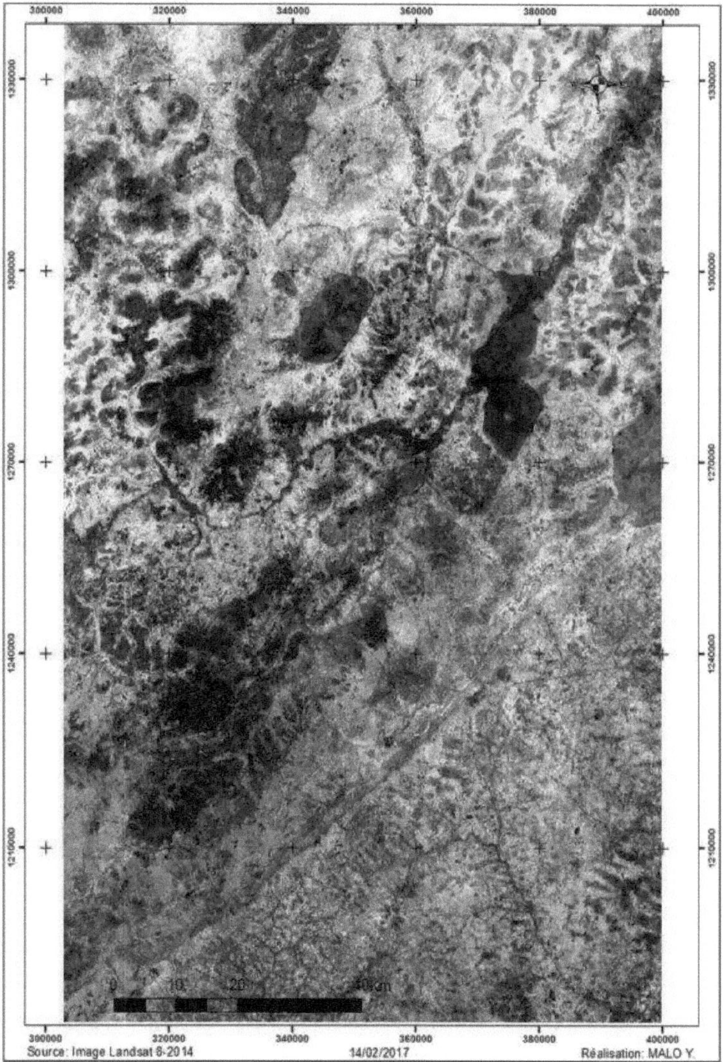

Map 14: The 7, 5, 3 colour composition of Landsat 8

It is possible to observe some features such as rocky outcrops in dark green, sedimentary formations in purple, water bodies in dark blue, latëritic formations. After the colour composition further processing was necessary to increase the image quality before classification.

4.2.3.2.   Image enhancement

This is a way of increasing the graphic quality of the image which can be done by band or by colour composition as in our study. The histogram presents a wide belly with pixels concentrated in the middle of the radiometric value axis. In this case, the image is medium grey (OUEDRAOGO B., 2015).

The goal was to reduce the minimum values to 0 and the maximum values to 255. To do this, we chose the **Equalization** option from the many image enhancement options in ENVI. Then, the image was filtered to remove, attenuate or accentuate spatial frequencies in order to sharpen linear themes or the contents of two or more contiguous themes for its proper interpretation (OUEDRAOGO B., 2015). The median filter [5 x 5] is chosen because it allows better detection of phenomena covering a large area, in particular the lithological contours of a given zone, large urban areas, oceans, in short homogenous themes.

In order to improve the final resolution of the image, we used Landsat 8 band 8 which has a resolution of 15m. The final composition now has a resolution of 15 m after the various processing operations.

Supervised classification is chosen for a customised identification of the different geomorphological units. The Region Of Interest (ROI) separability tool presented an optimal separability. The different geological formations and their residual reliefs showed almost no separability. This shows that these units have almost the same spectral signature and that they are composed of the same basic material (sandstone). This did not allow us to go beyond a number of geomorphological units. However, by merging and eliminating some geomorphic units we have achieved a satisfactory separability between the different geomorphic units and the pairwise separability values of the units are greater than or equal to 1.99.

4.3. The different geomorphological units

This is the set of units resulting from the classification of the landsat8 image with the description of the legend.

4.3.1. The geomorphological map

Several geomorphological units have been identified as a result of these different treatments. The map below presents these geomorphological units, of which there are more than ten, which can be divided into large groups, namely the plateau, the plain and the glacis. The details are very important and the units that at first sight seemed insignificant have turned out to be very important. The geomorphological map at 1:500,000 and the geological map have allowed us to correct the shortcomings.

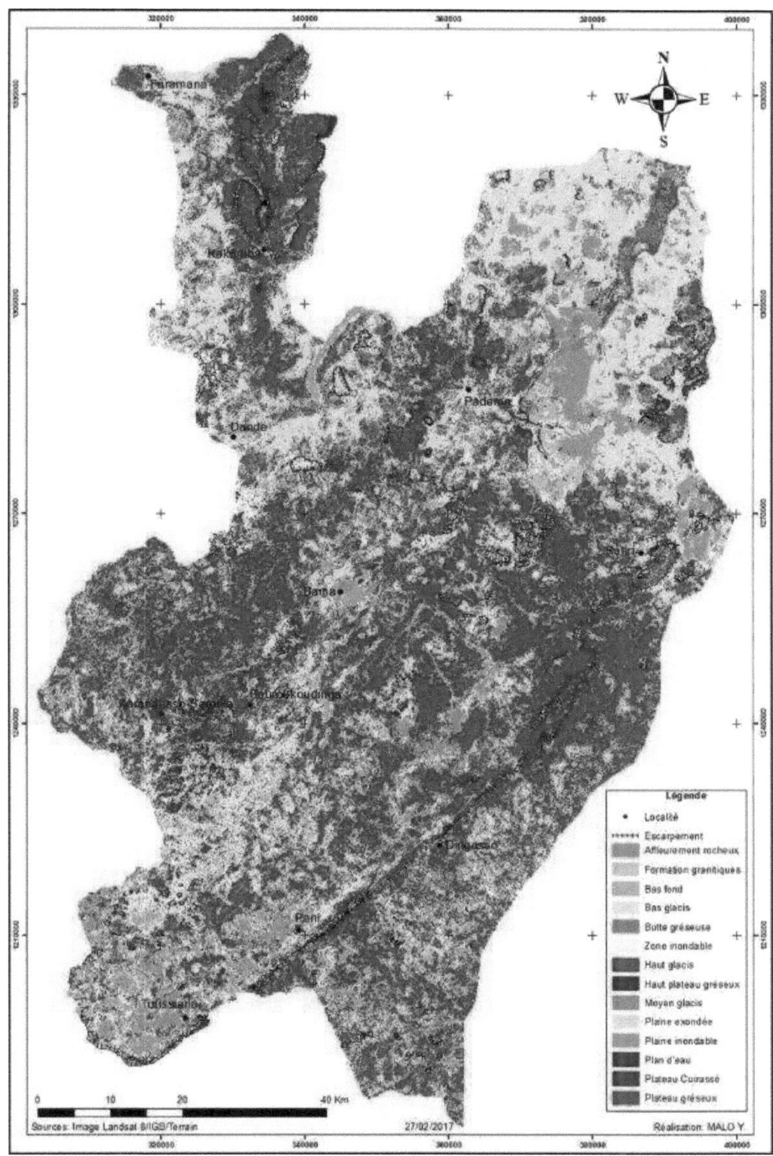

Map 15: Geomorphological units of the sandstone plateau of the Houet province

4.3.2. The legend

These are the different gëomorphological units represented on the map. These geomorphological units are proportionally distributed according to their respective areas in Table 5 below. With the exception of the crystalline formations, all the surface formations are underlain by sedimentary cover.

Table 7: The proportions of geomorphological units

Units geomorphological	Area (Km2)	Units geomorphological	Area (Km)^2
Glassy plateau	2422,13	Granitic formation	266
Armour plate	40,277	Rocky outcrop	292
Water body	1,168	Expanded plain	475,22
Low glaze	1035,89	Bottom	652,07
Medium glaze	12,06	High plateau	96,17
High glaze	362,33	Flood zone	1849,34
Flood plain	465,37	Gresous hillock	129,57

Source: Lanfsat image processing 8

4.3.2.1. The plain

It is subdivided into two types of plains: the floodplain and the floodplain. However, in the field we were able to observe the reclaimed plains of Bama and Takaledougou 1 and 2. In these plains, the valleys are wide and the rivers are not very deep. In the Houet, they form the transition between the sandy plateaus or between the sandy and the clay plateaus. Most villages are located on the plains. Fo, for example, is located in the depression between the butte temoin to the west and the commanding plateau. The rivers all flow eastwards apart from those that feed the Banifin. This flatness is often interrupted by residual reliefs, which can be buttes, tabular or rounded tops. This type of relief is located east of Toussiana, towards Padema and in the Koundougou region. These rivers are heavily silted up due to the origin of the water that is drained into them. The water flows from the elevations to the depressions.

Photo 2: The mound and the Fo depression.

Alluvial plains along the rivers exist in the study area. They are very important and occupy 21% of the surface area of the sandstone formations that we had qualified as floodplain.

4.3.2.2. The trays

These are the most important reliefs of the Houet province. Two types of plateaus have been identified: the sandstone plateaus and the armour plateaus. The reliefs of these plateaus are structural and dominated by layers of sandstone with various facies (hard and soft sandstone) whose inclination of the layers at the edge favours the establishment of a monoclinal structure.

4.3.2.2.1. The Giant Plateaux

They are flat surfaces inclined in one direction only, cut by incised valleys (DA D. E. C., 2012). In the sandstone plateau, the sedimentary layers are affected by a dip. They occupy a large part of the province and are visible over 32% of its area. They are interrupted by an escarpment (cliff) of more than 100 m high which separates them from a plain made up of much softer materials (the basal sandstone) in the case of the Banfora cliffs. This cliff occupies the entire southern and eastern part of the province and is about 100 km long. In its northern extension, it rests on the basement. To the north, another escarpment пошшё cliff of Fo-Bandiagara dominates a landscape that is sandy but covered by latëritic formations. This cliff is also ëdifying but more dissëquëe than that of Banfora. To the north-east, towards Padema, plateaus have formed on the Bonvate gres, a locality in the commune of Padema.

4.3.2.2.2. The Grese Highlands

These are high reliefs formed by sandstone. They dominate the other sandy plateaus by more than 100 m and represent 1.13% of the total geomorphic units. These plateaus are more represented in the southwestern part of the province.

4.3.2.2.3. The armour plateaus

They are less important in the province, namely 0.6% of all gëomorphological un^s. They occupy the east, southwest and northwest of the province. These plateaus, as is the case west of Dandë and Samandeni, consist of a succession of hills and armoured buttes. The slopes are often concave or convex depending on whether one is on the eastern, southern, northern or western flank. However, to the north of the Samandeni dam site, they begin with shale plateaus whose convex northern flank and concave western flank are occupied by angular, lamellar, south-north facing blocks.

4.3.2.3. The mounds

These buttes are residual reliefs but, composed of sandstone and separated from their commanding relief by dëpressions and plains. They may be lesser buttes, or foothills. These residual reliefs can be seen in the extreme east of Toussiana and between the granite chaos of Koro and the cuesta front of the northern part of the Banfora cliff. Other mounds can be found to the west of Kokoroba and Fo. The mounds occupy 2.69% of the area of the sandstone formations.

On about 35% of the territory area can be observed the sedimentary formations which are not covered by the surface materials. Some of the sandy formations are shown in the pictures 3 below.

Photo 3: A and B - The sandstone formations
Photo A: A sandstone mound east of the northern extension of the Banfora cliff. It is subject to differential erosion and some fragments of almost dismantled hard layers remain.

Cliché : MALO Y.

Photo B: Western flank of the Fo cliff. It is possible to observe fine to coarse sandstone. Some   pebbles Pudding quartzites are also visible.

Cliché : MALO Y.

4.3.2.4.   Granitic outcrops

It is therefore not surprising to talk about this type of formation, since we are on the edge of the Taoudeni sedimentary basin, which acts as a transition zone between the sedimentary and the crystalline. Moreover, the sedimentary formations rest directly on this base. The contact between these two formations, i.e. the lower sandstones which form the main part of the Banfora cliff, is quite clear at Borodougou on the eastern flank of a sandy hillock on which a sand quarry is in operation. It is a granite chaos which develops from the village of Koro and continues towards the village of Borodougou, leaving Yegueresso and Namandougou to the east, before continuing west of Kotedougou. The granite blocks are stacked from centimetre to metre size. It is a multi-facies granite according to HUGOT G. (2002). They are grey granitic facies and pink speckled granitic facies. Photograph 4 shows the grey and red granites.

Photo 4: The granitic and coarse red sandstone blocks of Borodougou

Cliche: YAMEOGO S. W. I. (2015)

The contact between the sedimentary cover and the basement is not only made in this part; it is also made in the vicinity of Koumi-Soundeni, Dingasso. To the west of Dande, the flanks of the cuirassed plateaus are paved in places with granitic blocks. On the geomorphological map, sandstone outcrops abundantly along some rivers and in depressions. The geological section realised south-west of Borodougou by YAMEOGO S. W. I. (2015), presents the stratigraphy of this contact.

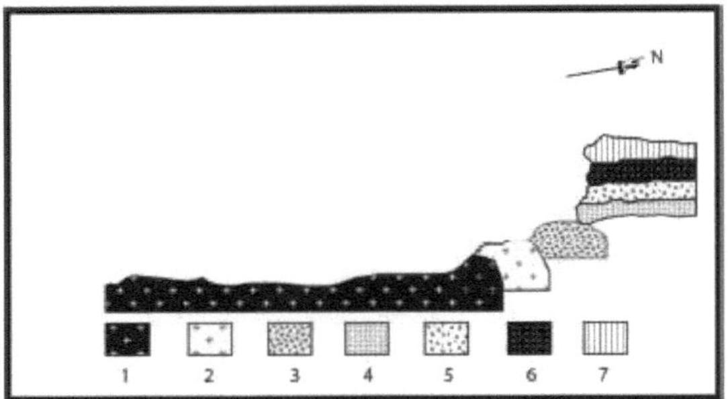

Figure 9: Geological section southwest of Borodougou (basement-sedimentary cover contact)

Source: YAMEOGO W. S. I. (2015).

Legend: bedrock (1: sound granite; 2: pink speckled granite; 3: red granite) sedimentary cover (4: fine gravel; 5: coarse conglomerated gravel; 6: coarse gravel; 7: fine interbedded gravel).

4.3.2.5.   Glazes

Glacis armour covers the underlying sandstone formations and those on sedimentary formations. According to DA D. E. C. (2012), glacis have a smooth, even slope and develop in soft rocks, old weathered granites, shales and friable gres in tropical regions at the foot of major landforms.

4.3.2.5.1.   The high glacis

It is the first level of the glacis and represents 4.4% of the study area. It is important in the centre, east and west of the province. In the north it marks a direct contact with the Fo cliffs. With an altitude often higher than 450 m. It ensures the transition between the plateaus and the middle glacis. It is sometimes covered with blocks of dismantled or dismantling armour. Their size varies from less than one cm in diameter to more than one metre in diameter.

4.3.2.5.2.   The medium glaze

It occupies the eastern and northern exit of Bobo and to the south 7 km from Darsalami. The middle glacis is composed partly of iron-formed armour and sandstone formations. The quartz-laterite system is very important in this formation, especially east of the town of Bobo-Dioulasso. This type of glacis is weakly represented (0.24%). It often lies directly on the sedimentary formations as is the case north of the town of Bobo-Dioulasso. To the north of Dande, it is covered with blocks of rubble of varying sizes according to our field observations. Its altitude is less than 450m and it provides a transition between the high glacis and the low glacis. The scree of dismantled armour, some of which has been used as a stone cordon, covers the concave eastern slopes of a middle glacis to the west of Dande. All these units are underlain by the underlying sedimentary formations.

Photo 5: The eastern slope of a medium glacis

Cliche: MALO Y.

4.3.2.5.3.   The low glacis

This is the low level (Figure 11) which provides the transition between the middle glacis and the depressions or plains. It represents 12.14% of the surface area of the sandy plateau of the Houet province. It is often observed a red clayey level with a medium degree of induration giving a scoriaceous aspect.

Figure 10: The topographic profile north of Dande

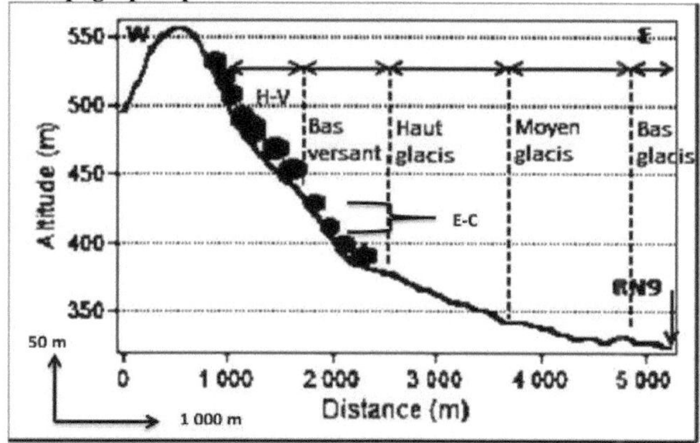

Source: SRTM/Terrain H-V: High slope; E-C: Scree

4.4. Drainage in the Grese Plateau

In order to understand the behaviour of the rivers draining the Houet province, a shading map was made. This map allowed to understand the orientation of the rivers and the general organisation of the hydrographic network of the Houet province.

71

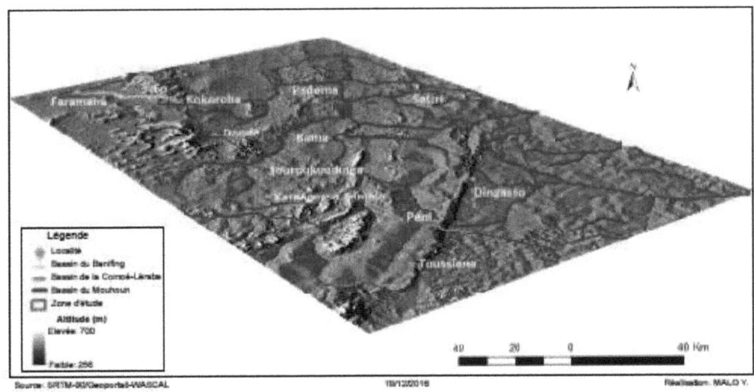

Map 16: Shading and organisation of the river system

On the shading map, the three basins that share the province are visible and the Mouhoun is the most important and drains most of the water. In the north, in the Fo cliffs, the dissection of the relief shows that there would have been an overlay as they are dissected. However, the rivers have several orientations in this part of the province. The cataclinal or consequential streams that feed the Niger Basin are oriented north-west, i.e. in the same direction as the general dip of the sedimentary formations in the province. They are opposed to the anaclinal or obsequential streams that flow south-eastwards to feed the Mouhoun basin.

To the east, north, north-east, east, west, centre, south-west and south, the waters are drained by the Mouhoun basin, especially its upper part. These are SW-NE trending streams, perpendicular to the general direction of dip. These rivers are therefore orthoclinal. But in this part, the dips present several orientations while being circumscribed. This is the case at Toussiana, Takaledougou, Kodima, Souroukoudinga, Bonvale where WNW, N, NNE, NW, SSW, SSE directions are observed with dips between 2 to 11° according to the geological sections made in these localities by OUEDRAOGO C. (1983). To the south-east at Dingasso, for example, the watercourses are anaclinal and oriented towards the south-east and drain the waters of the lower Mouhoun via the Koba tributary of the Bougouriba.

Finally, the Como®-L®raba basin is fed by orthoclinal streams that originate west of P®ni, perpendicular to the g®n®ral dip. It is a south-western direction that continues into the Сотоё province.

The map also shows the many furrows that cut into the slopes of the plateaus, hills and knolls before feeding the secondary and primary rivers of the province.

Partial conclusion

Satellite and radar imagery allowed us to produce geomorphological and geform maps of the sandstone plateaus of the Houet province. Fourteen gëomorphological units could be identified at this scale in the study area. The most important of these is the geological plateau which occupies about 35% of the total area. In some parts of the province, the contact between the sandstone formations and the basement is visible. These sandstone coverings are covered at certain levels by lateritic formations. This is what hides a large part of the gres. The plains are also abundant and two of them have been developed (Bama and Takaledougou) and another one is in progress. This is the Samandeni plain which extends into the Kenedougou province.

GENERAL CONCLUSION

This study on geomorphology, particularly the cartographic aspect, in the sandstone plateaus of the Houet province was a means of understanding the geomorphological units in this part of Burkina Faso. Beforehand, a lot of documentary research helped us to understand the geological formations of the study area because they are very important in the understanding of the morphology of a region. Thus, nine sedimentary formations exist in the province with a representation of the crystalline basement which is not taken into account in this study.

The different geoforms have been analysed, i.e. altitudes, slopes, lineaments. The high altitudes are found in the southwest and the low altitudes in the north and northeast. The average altitude in these sandstone formations is 452 m. The slopes vary between 0 and 37°. Thus, in some areas the slopes are zero or low and in others they are high. The lineaments are high in the depressions and low in the plateau regions. Following this, we say that our first hypothesis about the geoforms of the Houet province is verified.

Our results also allowed us to identify 14 geomorphological units with very significant sandstone escarpments in some areas. The major geomorphological units in the province include plains, plateaus (sandstone, limestone and shale) and glacis. The sandstone plateaus are the most important of all geomorphological units in the study area, occupying about 32% of its surface. From these results, we can conclude that our second hypothesis that several geomorphological units would be revealed is verified.

However, this study is far from having revealed all the geomorphological units of the sedimentary formations of the Houet province. Some details have yet to be provided as our current resources have not allowed us to go beyond this. Secondly, we have deliberately omitted the links between the geomorphology and the different elements of the physical environment and certain anthropic developments. These include the river system, soils, vegetation and climate. All these issues could be addressed in our subsequent studies on the same theme but in a much broader and more inclusive framework.

BIBLIOGRAPHY

AVENARD J. M., 1977: Notice explicative No 71 cartographie gëomorphologique dans l'ouest de la Cote d'Ivoire. ORSTOM, Paris, 109 pages.

BILODEAU C., 2010: Apports du lidar a l'etude de la vëgëtation des marais sates de la baie du Mont-Saint-Michel. Thëse de doctorat en Sciences de l'Information Gëographique. Universite Paris-Est, 210 pages.

BOEGLIN J.-L., 1990: Evolutions mineralogique et gëochimique des cuirasses ferrugineuses de la region de Gaoua (Burkina Faso). These-Universite Louis Pasteur, 188 pages.

CLENET H., 2009 : Teledetection hyperspectrale : mineralogie et petrologie, Application au volcan Syrtis Major (Mars) et a l'ophiolite d'Oman. These, Universite Toulouse III - Paul Sabatier, 367 pages.

DA D. E. C., 1984: Recherches geomorphologiques dans le sud-ouest de la Haute-Volta. Dynamique actuelle en Pays Lobi. These de doctorat de troisieme cycle de geographie. Universite Louis Pasteur de Strasbourg, 309 pages.

DAVEAU S., 1960: Les plateaux du sud-ouest de la Haute-Volta: Etude geomorphologique. Faculte des Lettres et Sciences Humaines de Dakar, 68 pages.

DEMANGEOT J., 1999: Tropicalite, Geographie physique intertropicale. Armand Colin, Paris, 340 pages.

DIDIER R. de S.-A., 1969: Le Continental Terminal et son influence sur la formation des sols au Niger. cah. O.R.S.T.O.M., &. Pedol, vol. VII, no 4, 24 pages.

<u>Direction Generale des Statistiques Animates, 2008</u>: Les statistiques du secteur de ltelevage au Burkina Faso, 124 pages.

EGGLETON R. A., 2001: The Regolith Glossary surficial geology, soils and landscapes. Cooperative Research Centre for Lanscape Evolution and Mineral Exploration, 151 pages.

GRIMAUD J.-L., 2014 : Dynamique long-terme de l'erosion en contexte cratonique : l'Afrique de l'Ouest depuis l'Eocene. These-Université Toulouse III Paul Sabatier-303 pages.

GUINKO S., 1984: Vëgëtation de la Haute Volta. Thëse, tome 1-Universite de Bordeaux III335 pages.

GUIRAUD R., 1988: L'hydrogëologie de I'Afrique. Journal of African Earth Sciences, 7, 3, 519-543.

GUYOT-SAVONNET C., 1986: Etat et sociëtës au Burkina Faso: Essai sur le politique africain. KARTHALA, Paris, 223 pages.

HERBERT J. 1969: "les Gwi et les Turko". Notes et documents voltai'ques 3 (1).

HUGOT G., 2002: A la recherche du Gondwana Perdu: Aux origines du monde. 311 pages.

Institut National de la Statistique et de la Demographie/Ministere de l'Economie et des Finances, 2010: La region des Hauts-Bassins en chiffres. 8 pages.

IWACO/Ministere de l'eau-B.urkina^aso/Direction Generale de la Cooperation au Developpement- Netherlands, 1993: Carte hydrogëologique du Burkina au 1/500 000 (Feuille de Bobo). 41 pages.

JOHAN D., 2006: Modëlisation des eaux souterraines. 35 pages plus appendices.

KALOGA B., 1969: Etude de la pëdogenëse sur les glacis soudaniens de Haute-Volta. Bull. Ass. sënëg. Et. Quatern. Ouest afr, Dakar, no 22, juri-? 1969 4 pages.

KOUSSOUBE Y., 2013: Hydrogëologie des sëries sëdimentaires de la dëpression piëzomëtrique du Gondo (bassin du Sourou): Burkina Faso / Mali, Hydrology, Universite Pierre et Marie Curie- Paris VI, 2010, 271 pages.

METELKA V., 2011: Geophysical and remote sensing methodologies applied to the analysis of regolith and geology in Burkina Faso, West Africa. Thesis-Prague, Toulouse- 207 pages.

MIETTON M., 1988: Dynamique de l'interface lithosphëre-atmosphëre au Burkina Faso, l'erosion en zone de savane; thèse de doctorat d'Etat, Universite de Grenoble; 511 pages.

Ministere des Forets, de la Faune et des Parcs /Secteur des forets, 2015 : Guide d'interprétation des mosaiques d'images satellitaires Landsat. Quebec, CANADA, 22 pages.

Ministry of Economy and Finance, 2010: Profil des régions du Burkina Faso. Ouagadougou-Burkina Faso, 456 pages.
Ministry of Economy and Finance, 2009: Monographie de la Rëgion des Hauts- Bassins. 154 pages.

NABA S., 2007: Propriëtës magnetiques et caracteres structuraux des granites du Burkina Faso oriental (Craton Ouest Africain, 2,2 - 2,0 Ga) : implications geodynamiques. These-Universite Toulouse III- Paul Sabatier-175 pages.

NAKOLENDOUSSE S., 1991: Methode d'évaluation de la productivite; des sites aquiferes au Burkina Faso: Geologie-Geophysique-Teledetection. These-Universite Joseph Fourier-Grenoble I- 258 pages.

OUATTARA G., 1998: Structure du batholite de Ferkessedougou (Secteurde Zuenoula, Cote d'Ivoire). Implication sur Interpretation de la Geodynamique du paleoproterozoique D'Afrique de l'Ouest A 2.1.Ga. These-University of Orleans-343 pages.

OUEDRAOGO C., 1983: Etude geologique des formations sedimentaires du bassin precambrien superieur et paleozoique de Taoudeni en Haute-Volta. These, Universite de Poitier, 210 pages.

OUEDRAOGO C., 2002: Carte geologique a 1/200 000^e du degre carre de Dedougou (IGB, ed.), BUMIGEB-Bobo, contrat-plan 2000-2004. BF.

OUEDRAOGO C., 2006: Programme de valorisation des ressources en eau de l'Ouest.

Synthese geologique de la region Ouest du Burkina Faso. Rapport definitif. FED. 46 pages.

OUEDRAOGO I., 1994: geology and hydrogeology of the sedimentary formations of the Boucle du Mouhoun (Burkina Faso). These- Faculte des Sciences et Techniques de Dakar (Senegal). 159 pages.

OUEDRAOGO I., 2010: Land Use Dynamics and Demographic Change in Southern Burkina Faso. Doctoral Thesis Swedish University of Agricultural Sciences (Alnarp). 64 pages.

PICOUET C., 1999: Geodynamique d'un hydrosysteme tropical peu anthropise: Le Bassin superieur du Niger et son delta interieur. These-Universite Montpelier II. 469 pages.

Programme de Valorisation des Ressources en Eau dans le Sud-ouest (RESO)/IWACO-BURGEAP, 1998: Diagnostic des ressources en eau dans le bassin du Mouhoun Superieur. 49 pages plus annexes.

SANOU D. C., 1984: Quelques problemes de dynamique actuelle: L'erosion des sols dans la region de Bobo-Dioulasso, Burkina Faso. These de doctorat de 3e cycle, Universite Louis Pasteur, Centre de geographie appliquéee, Strasbourg, 212 pages + annexes.

SANOU D. C., 2008: ABC de la geomorphologie structurale (cours et T.D.) 107 pages.

SAVADOGO A. N., 1984: Geologie et hydrogeologie du socle cristallin de Haute Volta: Etude regionale du Bassin Versant de la Sisssili. These-Universite Scientifique et Medicale de Grenoble (France). 179 pages

SAVADOGO A. N., 2012: Catchment fields on megafractures of the basement and generalization of drinking water supply networks in rural areas in Burkina Faso. 12 pages.

YAMEOGO S., 2008: Ressources en eau souterraine du centre urbain de Ouagadougou au Burkina Faso *Qualite et vulnerabilite*. These- University of Avignon and Pays de Vaucluse-254 pages.

TCHALARE B. M., 2015: Evolution geomorphologique plio-quaternaire et dynamique actuelle dans le bassin de Sotouboua (Region centrale du Togo). These-Université de Ouagadougou- 341 pages.

THOMAS Y. F., 1972: Cartographie des fonds sedimentaires littoraux - projet de legende. Centre National pour l'Exploitation des Oceans-38 pages.

MASTERS AND MASTERS' THESES

DAKOURE D., 1999 : Composition isotopique des precipitations a Bobo-Dioulasso et relation avec les eaux souterraines des différents unites des formations sedimentaires et du socle du sud-ouest du burkina faso. Universite de Paris Sud-DEA-50 pages.

ILBOUDO S., 2010: Contribution de la geomatique a la recherche miniere et a la préservation de l'environnement sur le site aurifere de Dossi. Memoire de Master-SIG-

AGEDD, Universite Ouaga I Professeur Joseph Ki ZERBO, 120 pages.

INQUA-ASEQUA SYMPOSIUM INTERNATIONAL, 1986: Global Change in Africa during the Quaternary: Past-Present-Future. Dakar, ORSTOM - 485 pages.

KABORE D., 2014: L'analyse du potentiel hydrogeologique de la Commune rurale de Tanghin-Dassouri (Province du Kadiogo). Master-ISESTEL-102 pages.

KONATE S., 1995 : les mosques dans la sociëtë Zara de Kawë (Lahirasso), Province du Houet. Memoire de maitrise- Department of History and Archeology. 117 pages.

LACAN C., 2010 : Analyse geomorphologique des bassins versants de l'Ogooue et du Congo : Quantification de l'erosion Cenozoi'que et implications sur les flux detritiques vers la marge d'Afrique Equatoriale. Master-Universite Montpellier II-32 pages.

NARIMENE I., 2012: Utilisation de la teledetection pour la cartographie geologique du Massif des Eglab et de sa bordure sedimentaire (Sud-Ouest algerien) : Exemple de la feuille de Mokrid. Master-UNIVERSITY FERHAT ABBAS - SETIF (Algeria) - 68 pages.

OUEDRAOGO H., 1992: La degradation du couvert végétal et ses conséquences socio-économiques dans la region de Toussiana (Province du Houet). Memoire de maitrise, University of Ouagadougou-86 pages.

SANOU D. C., 1981: Etude comparative entre une parcelle pourvue de bourrelets anti-erosif et des parcelles traditionnelles a Sirgui (Kaya); Introduction aux problemes de dynamique erosive, Memoire de maitrise, Université de Ouagadougou, 102 pages.

SOULAMA K., 2016 : Implantation et évolution du Wahhabisme dans la region des Hauts-Bassins : les cas de Bobo-Dioulasso, de Dieri, de Hounde et de Orodara, de 1938 et 2013. Master's thesis II UO, Department of History and Archeology. 169 pages.

YAMEOGO W. S. I., 2015 : Caracterisation de l'alteration albitisante au contact socle cristallin / couverture sedimentaire du secteur de Koro - Kotedougou (Burkina Faso, West Africa). Memoire de Master U O/ Departement des Sciences de la Terre.

ZOUNDI N., 2001: Les stratégies de lutte contre le paludisme et itineraires therapeutiques concernant les enfants de moins de cinq ans: une etude comparative en milieu urbain et rural dans la province du Houet. Memoire de maitrise, University of Ouagadougou, 96 pages.

<u>ARTICLES</u>

ARWYN J. et al, 2015: Atlas of the soils of Africa. 178 pages.

AUCOUR A.-M., CARBONEL P., FABRE J. and PETIT-M., 1986: Sëdimentation lacustre Holocene de la region de Taoudeni page 9-12. INQUA/ DAKAR SYMPOSIUM "Changement globaux en Afrique".

BEAUDET G., COQUE R. 1994: Reliefs et modèles des regions tropicales humides: mythes, faits et hypotheses. In: Annales de Geographie. t. 103, n°577. pp. 227-254.

BETARD F. and BOURGEON G., 2009: Cartographie morphopedologique: de revaluation des terres a la recherche en geomorphologie. (Geomorphologie : relief, processus, environnement, 2009, n°3, p. 187-198). 13 pages.

CHABREUIL M. and A., 1979: Exploration de la terre par les satellites. Hachette, 155 pages.

CHEREl J.-P., 2010: Support de cours M1 SIIG3T - Traitement d'images / transformation d'images de teledetection. 4 pages.

DA D. E. C., 1989: Exploitation des imageries satellitaires Landsat TM pour la cartographie geomorphologique dans le centre-nord du Burkina Faso. CAHIERS DU CERLESHS, N°4 mois 05. 31 pages.

DA D. E. C., 2005: GIS and geomorphological mapping at 1:500,000 of Burkina Faso. 15 pages.

DA D. E. C., YACOUBA H. 2 and YONKEU S. 2, 2008: Unites morphopedologiques et gestion de la fertilite des sols dans le Centre-Nord du Burkina Faso par les populations locales. 12 pages.

DOSSIN F., REKK S., SOREL A., HALLET V., 2009: Notice explicative de la carte hydrogeologique de Wallonie. Facultes universitaires Notre Dame de la Paix de Namur (Belgium). 144 pages.

FONTES J. and GUINKO S., 1995: Carte de la vegetation et de l'occupation du sol du Burkina Faso. Notice explicative. Institut du Developpement Rural, Faculte des Sciences et Techniques- Universite de Ouagadougou (Burkina Faso) 71 pages.

GAUDIN S., 1997: Quelques elements de geologie. BTSA Gestion forestiere- 31 pages.

GODARD A. and TABEAU M., 2009: Les climats : Mecanisme, variabilite, repartition. Armand Colin, Paris, 216 pages. *

GEORGE P. and VERGER F., 2009: Dictionnaire de la geographie. France Query-ZA. des Grands Camps. 480 pages.

HOTTIN G., OUEDRAOGO O. F., 1975: Notice explicative de la carte geologique a 1/1000 000^e de la republique de Haute-Volta. Editions du BRGM, Paris, 58 p.

HUYSECOM E., JEANBOURQUIN C., MAYOR A., CHEVRIER B., LOUKOU S., CANETTI M., DIALLO M., BOCOUM H., GUEYE N S., HAJDAS I., LESPEZ L. & RASSE M., 2013 : Reconnaissance in the Faleme Valley (Eastern Senegal): the 15th year of research of the international programme "Peuplement humain et paleoenvironnement en Afrique de l'Ouest". Universite de Geneve 89 pages.

ILBOUDO H., LOMPO M., WENMENGA U., NABA S., KAGAMBEGA N., TRAORE S. A., 2008 : Caracteres petrographiques et structuraux des formations Paleoproterozoiques

(Birimien) du gite a sulfures de Tiebele, Burkina Faso (Afrique de l'Ouest). Journal des Sciences, 14 pages.

JEAN-CLAUDE D., 1965: TRICART J., Principes et methodes de la geomorphologie. Paris, Masson, 1965. 496 p. Cahiers de geographie du Quebec, vol. 10, n° 19, 1965,3p.- p. 150-151.

MALLET C., LAFON V., DESPRATS J. F. 2007: Cartographie des facies geomorphologiques du littoral aquitaine a partir des données FORMASAT-2. BRGM/RP-56101-EN. Final report-69p., 39ill.

MASSON M., 1972: Utilisation de la geomorphologie dans les etudes geotechniques de sites a urbaniser. Geologue-Laboratoire regional de Rouen-15 pages.

OKAINGNI J.-C., KOUAME K. F., MARTIN A., 2010: Cartographie des cuirasses dans les formations volcanosedimentaires de la zone d'Anikro-kadiokro (cote d'ivoire) a l'aide de la théorie des fonctions de croyance. *Revue Teledetection*, 2010, vol. 9, n 1, p. 19-32-14pages.

OUATTARA G., KOFFI G. B. and YAO A. K., 2012 : Contribution des images satellitales Landsat 7 ETM+ a la cartographie lithostructurale du Centre-Est de la Cote d'Ivoire (Afrique de l'Ouest). *International Journal of Innovation and Applied Studies - ISSN 2028-9324 Vol. 1 No. 1 Nov. 2012, pp. 61-75.*

OUEDRAOGO B., 2015: Teledetection: digital image processing under ENVI. Course, 64 pages.

PETIT M., DA E. C. and GRANDIN G., 1994: Carte geomorphologique du Burkina Faso au 1/1.000.000 33 pages.

SANOU D. C., 1998: Dynamique superficielle et lutte anti-erosive a Zëcko (Vltinev Burkina Faso, Universite de Ouagadougou. Project/CRISAT, DMN, 108 pages).

SANOU A. and BARY A., 2012: Le marigot horUt a Bobo-Dioulasso: une question de sante publique? 13 pages. International Journal of Biological and Chemical Sciences.

WADE S., RUDANT J.-P., BA K., NDOYE B., 2007: Teledetection, GIS and geohazards: applications to the study of urban flooding in Saint-Louis and gullying due to water erosion in Nioro-du-Rip (Senegal). *Proceedings of the JSIRAUF, Hanoi, 6-9 November 2007*

YONKEU S., 2006: Evaluation environnementale stratégique du programme de securite alimentaire et nutrition, de fertilite des sols du Burkina Faso. ETSHER Group, Burkina Faso-12 pages.

WEB SITES

Part Three - The geomorphological map: http://geographie2001.free.fr/TDgeo.html (24/08/2016).

https://fr.vikidia.org/wiki/Bassin sedimentair (10/09/2016).

Roger COQUE, "**BASSIN SEDIMENTAIRE**", *Encyclopedia Universalis* [online], accessed on URL : http://www.universalis.fr/encyclopedie/bassin-sedimentaire/ 10/09/2016.

http://littre.reverso.net/dictionnaire-francais/definition/gres/35776 10/09/2016.

http://geographie2001.free.fr/TDgeo.html 24/08/2016

Driss Chahid, Larbi Boudad, Abdellah El Hmaidi, Ali Essahlaoui, Arnaud Lenoble, 2012 : Contribution of GIS and DTM in the study of the barrier beaches of the Temara region (Morocco). https://halshs.archives-ouvertes.fr/halshs-00714522 11/11/2016.

https://apropos.erudit.org/fr/usagers/politique-dutilisation/
Ph. Leveau, Burno, *Rivers and marshes, a history at the crossroads of nature and culture,* Comite des travaux historiques et scientifiques, 2004. 11/07/2017
https://fr.wikipedia.org/wiki/Cours of water 11/07/2017
https://www.aquaportail.com/definition-3804-cours-d-eau.html 11/07/2017
https://fr.wikipedia.org/wiki/Vall%C3%A9e 11/07/2017

Printed by Books on Demand GmbH, Norderstedt / Germany